NOT HERE, NOT THERE,
NOT ANYWHERE

NOT HERE, NOT THERE, NOT ANYWHERE

Politics, Social Movements, and the Disposal of Low-Level Radioactive Waste

DANIEL J. SHERMAN

Washington, DC • London

First published in 2011 by RFF Press, an imprint of Earthscan

Earthscan LLC, 1616 P Street, NW, Washington, DC 20036, USA
Earthscan Ltd, Dunstan House, 14a St Cross Street, London EC1N 8XA, UK
Earthscan publishes in association with the International Institute for Environment and
Development

For more information on RFF Press and Earthscan publications, see www.rffpress.org and
www.earthscan.co.uk or write to earthinfo@earthscan.co.uk

ISBN: 978-1-933115-91-7 (hardback)
ISBN: 978-1-933115-92-4 (paperback)

Copyedited by Joyce Bond
Typeset by OKS Press Services
Cover design by Circle Graphics

Library of Congress Cataloging-in-Publication Data

Sherman, Daniel J.
  Not here, not there, not anywhere : politics, social movements, and the disposal of low-level radioactive
waste / Daniel J. Sherman.
      p. cm.
    Includes bibliographical references and index.
    ISBN 978-1-933115-91-7 (hardback : alk. paper) – ISBN 978-1-933115-92-4
    (pbk. : alk. paper)
    1. Radioactive waste disposal–United States.
  I. Title.
  TD898.118.S44 2010
  363.72'89–dc22   2010024625

A catalogue record for this book is available from the British Library.

About Resources for the Future *and* RFF Press

Resources for the Future (RFF) improves environmental and natural resource policymaking worldwide through independent social science research of the highest caliber. Founded in 1952, RFF pioneered the application of economics as a tool for developing more effective policy about the use and conservation of natural resources. Its scholars continue to employ social science methods to analyze critical issues concerning pollution control, energy policy, land and water use, hazardous waste, climate change, biodiversity, and the environmental challenges of developing countries.

RFF Press supports the mission of RFF by publishing book-length works that present a broad range of approaches to the study of natural resources and the environment. Its authors and editors include RFF staff, researchers from the larger academic and policy communities, and journalists. Audiences for publications by RFF Press include all of the participants in the policymaking process—scholars, the media, advocacy groups, NGOs, professionals in business and government, and the public. RFF Press is an imprint of **Earthscan**, a global publisher of books and journals about the environment and sustainable development.

CONTENTS

LIST OF FIGURES AND TABLES

ACKNOWLEDGMENTS

I am grateful to have benefited from the critical guidance and patient support of many people over the course of this project. I continue to enjoy the love, support, and brilliant (if brutally honest) insights of Christina Sherman, to whom this book is dedicated. Audrey and William—who were not yet born when this project began—have helped me keep my work in perspective. I am grateful to have loving and supportive parents, Sally and John, who even went so far as to help me hunt microfilm in North Carolina and Nebraska. This project would not have been possible without the intellectual contributions of my mentors Sidney Tarrow, Theodore Lowi, and Walter Mebane. My work was supported by the Morris K. Udall Foundation, Teresa Heinz Scholars for the Environment Program, National Science Foundation, Henry Luce Foundation, and Martin Nelson Junior Sabbatical Program. I am especially thankful to my colleagues in the Department of Politics and Government and the Environmental Policy and Decision Making Program at the University of Puget Sound for creating an ideal atmosphere for lifelong teaching and learning. Thanks to Don Reisman and his team at RFF Press as well as three anonymous reviewers for their helpful advice and constructive feedback during the publication process. Finally, I owe a debt of tremendous gratitude to the kind people who agreed to share their stories and experiences during interviews.

— *Daniel J. Sherman*

To Christina, my true love

CHAPTER 1

NOT HERE, NOT THERE, NOT ANYWHERE: AN INTRODUCTION

In the summer of 2008, newspapers in Pennsylvania reported that nuclear power plants, research facilities, hospitals, and other commercial producers of low-level radioactive waste (LLRW) in the state could no longer ship their contaminated by-products to South Carolina for disposal. Some of the LLRW materials with the lowest and shortest-lived radioactivity, like most of those generated from medical procedures, could be managed on-site until stable enough to handle safely or shipped more than 2,000 miles away for final disposal at a facility in Tooele County, Utah. But other materials with higher, riskier, and longer-lived radioactivity, such as contaminated resins, tools, and parts from the Three Mile Island nuclear power plant near Harrisburg, now had no available disposal options. Commercial generators of LLRW in 35 other states also found themselves in this position.

This news mirrored stories nearly 30 years prior, when the state of South Carolina turned back a shipment of LLRW consisting of materials from the near core meltdown of the Three Mile Island power plant in 1979. Washington and Nevada, the other two host states of LLRW disposal facilities, soon joined South Carolina in restricting disposal access and demanding a new federal policy governing radioactive waste. The resulting Low-Level Radioactive Waste Policy Act (LLRWPA) of 1980 and its subsequent amendments were designed to provide a multitude of new LLRW disposal sites equitably distributed across the United States. In fact, Pennsylvania has been joined for the past 24 years under this law with Delaware, Maryland, and West Virginia in the Appalachian Compact and charged with establishing a regional LLRW facility. Although the Appalachian Compact continues to exist as a legal entity, its office was closed and its operations were terminated in 1998. No regional compact or state has successfully established a new site to receive LLRW under the LLRWPA—yet the law remains.

In 1979, the restrictions on LLRW disposal caused panic among LLRW generators, state governments, and Congress. Representative Morris K. Udall, from Arizona, argued that without passage of the LLRWPA, "we would be faced with the prospect of our hospitals and research institutions, as well as commercial nuclear power facilities, being forced to shut down" (*Congressional Record* 1980a, 34130). By the mid-1980s, states and regional compacts had engaged in a flurry of site selection processes for new LLRW facilities under the LLRWPA. The new law ushered in what the U.S. Department of Energy (DOE) called the "era of state responsibility" for the disposal of LLRW. In fact both the National Governors Association (NGA) and National Conference of State Legislatures (NCSL) had actively lobbied Congress for state control over disposal of commercially generated LLRW. Congress took the unprecedented step of devolving responsibility for the disposal of commercially generated LLRW to state governments and regional compacts by passing the LLRWPA. Between 1986 and 1993, individual states and regional compacts identified more than 20 U.S. counties as candidate sites for LLRW disposal facilities.

"Not here, not there, not anywhere!" was one of the slogans chanted by opponents to proposed LLRW disposal facilities in these candidate counties across the United States. Activists in these counties organized more than 900 events of collective public opposition, ranging from petition drives to clashes with state troopers by masked protesters on horseback. By 1999, the U.S. General Accounting Office (GAO) reported that "states, acting alone or in compacts, had collectively spent almost $600 million attempting to develop new [LLRW] disposal facilities. However, none of these efforts have been successful" (GAO 1999, 3). The opponents' slogan seemed to be a reality.

This is a book about the implementation vulnerabilities of devolution and the political power of local collective action in the U.S. federal system. It explores the implementation vulnerabilities of the LLRWPA as the politics of LLRW disposal cycled through federal, state, and local levels of government. This political saga provides the ideal context from which to engage fundamental policy questions concerning the disposal of environmentally undesirable materials specifically, as well as the dynamics of local collective action and intergovernmental relations more generally.

An examination of implementation efforts under the LLRWPA reveals the power of local government and citizen opposition to thwart national policy objectives. Although states have preemptive authority over local governments, such as the candidate counties for LLRW sites, state officials lack the political will to implement the LLRWPA in the face of local opposition. Neither top-down nor bottom-up "voluntary" approaches to the siting process have successfully established new LLRW sites. Although active opposition was not constant across the candidate counties, and some states made more implementation progress than others, early and intense local opposition in some states altered the implementation calculus nationwide. Activists and local elected officials in many candidate counties effectively persuaded state officials to halt the search for LLRW sites. As implementation failed in these states, even states with candidate counties that were not mounting active opposition stopped their implementation efforts. This was due to a fear that the first newly created LLRW facility would not merely be one of many regional sites, but instead would become a national site and bear a disproportionate amount of the risks associated with LLRW disposal.

Yet despite the title of this book and the desires of LLRW opponents, the classical physics principle on the conservation of matter dictates that as long as the waste is generated, it must be going somewhere. And indeed, LLRW-producing activities continue in the United States, and the waste *is* going somewhere. Just as remarkable as the panic that ensued when LLRW disposal capacity was constricted in 1979 was the relative calm that greeted the 2008 closure of the Barnwell, South Carolina, LLRW facility. Although this closure left just two remaining facilities, it caused no panic and led to no response from lawmakers at the state or federal level. When asked if the Appalachian Compact would restart its search for a new LLRW site, the chief of nuclear safety at the Department of Environmental Protection in Pennsylvania, one of the largest producers of commercial LLRW, explained that "economically, it wouldn't make sense because the volume of trash stranded by the Barnwell closing is too small" (Lenton 2008, A1).

An unusual and unanticipated equilibrium took form under the LLRWPA even as the law failed in its stated purpose to establish new disposal sites equitably distributed across the states. The scarcity of disposal options led waste generators to pursue innovative waste reduction strategies. Dramatic reductions in waste volume eliminated the need for the multitude of new facilities the LLRWPA was designed to establish. At the same time, the authority granted to states and regional compacts under the LLRWPA enabled revenue-generating surcharges, disposal fees, and waste importation controls that made the continued acceptance of LLRW at existing sites attractive to host states. For the time being, commercial LLRW generators have succeeded in cobbling together an adequate combination of waste reduction measures, on-site interim storage, and limited disposal availability, even while local LLRW opponents and antinuclear activists have succeeded in thwarting the construction of regional LLRW facilities at fresh sites under the LLRWPA. Whether this unusual equilibrium remains in balance depends in large part on the continued survival of the devolutionary policy context created by the LLRWPA.

The 30 years of implementation efforts under the LLRWPA across local, state, and regional government units raise significant questions. First, in terms of policy formation, why did Congress devolve responsibility for commercial LLRW disposal to the state level? Why did states ask for this responsibility? And more remarkably, why was no consideration given to the possibility of local opposition? Second, in terms of implementation, how were the candidate sites for LLRW facilities chosen? What effect did the selection processes have on the eventual local opposition? Third, in terms of collective action and social movements, what determines the collective mobilization of local opposition? Why do different locales use different collective tactics of opposition? And what effects do varying levels of mobilization and varying tactics have on implementation? Finally, what does the interplay between devolutionary policy and local opposition imply for intergovernmental relations? How stable is the current policy context for LLRW generation and disposal?

These questions are relevant to policy contexts broader than merely the disposal of LLRW. Nuclear power, an energy source producing negligible greenhouse gas emissions, is currently enjoying a renaissance in the United States as a perceived solution to global warming. After nearly 30 years without any applications for new nuclear power plants in the United States, utilities have submitted 26 applications for new reactors. Both presidential candidates in the 2008 election endorsed the expansion of nuclear power as an energy solution. Yet neither the United States nor any other nation has successfully built a permanent waste site for radioactive waste. Yucca Mountain, Nevada, had been the preferred candidate site for high-level radioactive waste in the United States for more than 20 years and had even been considered an option for the disposal of LLRW after the LLRWPA failed to produce any new sites. In 2009, President Barack Obama and Congress effectively removed the site from consideration by defunding the project.

The determination of any new site selection process for radioactive waste from a new generation of nuclear power plants should be informed by the political dynamics of intergovernmental relations and social dynamics of mobilized local opposition that the failed LLRW siting attempts of the 1980s and 1990s brought to light. The current waste management balance for LLRW should also inform how decisionmakers and stakeholders think about the relationship between waste-producing activities and the availability of disposal. Even other energy sources that might be considered alternative solutions to global-warming and energy security problems and do not share the waste disposal problems of nuclear power, such as wind and solar, have had to engage implementation challenges during site selection reminiscent of the efforts to find LLRW disposal sites. With the passage of decentralized state policies dictating portfolio standards for alternative energy, utilities have faced implementation struggles to site both wind farms and large-scale solar projects. These struggles, such as the Cape Wind project in Massachusetts and the BrightSource solar project in the Mojave Desert, have included mobilization of local opposition and contentious relationships among local, state, and federal government actors. The BrightSource project shares both geography and an intergovernmental constellation of opposition forces with a 1990s candidate site for LLRW disposal.

A BRIEF SUMMARY

Each chapter of this book grapples with one or more the above questions. Chapter 2 examines more than 50 years of nuclear energy legislation, congressional hearings, debates, and other documents to clarify the context in which Congress decided to devolve responsibility for LLRW disposal to the states. This was a remarkable policy change in that radioactive waste policy has been historically dominated by a closed loop of technical experts within the federal bureaucracy. The fact that the NGA and NCSL pushed for the devolution of waste-siting responsibility is even more startling. The policy history reveals a federal nuclear bureaucracy that steadily lost environmental and safety credibility in the eyes of state governments. This trajectory of distrust had roots in the very design of the Atomic Energy Act of 1954 and the dual roles assigned to the Atomic Energy Commission (AEC) to at once promote commercial nuclear power and regulate the emerging industry. The agency pursued the former role at the expense of the latter, which led to the embarrassing risks to health and safety that caught the attention of both the media and other political actors, including state governments.

Throughout the 1970s, the states had been demanding greater decisionmaking authority over nuclear facilities operating within their boundaries. This state encroachment into the regulation of nuclear policy developed in response to the revelation of federal regulatory shortcomings during the 1960s and 1970s. The near meltdown of the Three Mile Island nuclear power plant in 1979 was the most infamous example of more than a decade of inadequate federal regulation. In 1980, the states actively lobbied the federal government for more decisionmaking authority over all types of radioactive waste disposal. The federal government devolved disposal responsibility of LLRW to states as a token, while it jealously guarded authority over high-level waste disposal. The policy history leading to passage of the LLRWPA contains important lessons about the cumulative and layered nature of political development and policy change, as well as the role subjective problem definitions play in the formation of policy solutions.

Chapter 3 examines the site selection processes that state siting authorities followed as they attempted to implement the LLRWPA. The design of the 1980 law was blind to two key intergovernmental dynamics: the pressure local politics can place on state political actors and the interstate dynamics of waste avoidance. Amendments to the LLRWPA in 1985 placed strict timelines on state siting processes and threatened harsh penalties for states that failed to advance an LLRW disposal solution for waste generated within their borders. This seemed to address the latter intergovernmental blind spot in the original law, and states began to identify candidate sites for LLRW facilities in 1986. By this time, the phenomenon of local opposition to waste siting was well known and even feared by state siting authorities to such an extent that it became part of the LLRW problem definition. There is evidence in many cases that concern regarding the potential power of local opposition led siting authorities to augment technical site selection criteria with demographic profiles designed to identify acquiescent communities. Relative ratio environmental justice analysis shows that an overwhelming majority of the candidate sites were located in counties with disproportionately high low-income populations. If those charged with identifying new LLRW sites did indeed intend to avoid local opposition, the effort failed. Content analysis of letters to the editor in local papers for every day of the siting processes across 21 candidate sites shows not only overwhelming opposition to the LLRW facilities, but also a strategic refusal of opponents to engage arguments typically associated with the NIMBY ("not in my backyard") frame of meaning—a frame that siting authorities often conjured to depict the public as "poorly informed, interested primarily in avoiding local imposition of risks, and emotive rather than cognitive in its appraisal of the risk and in its response to siting proposals" (Kraft and Clary 1993, 96). Instead, opposition activists deftly juxtaposed the technical criteria established to guide site selection with the socioeconomic and political aspects of the siting processes employed by siting authorities—creating a sophisticated injustice frame that attacked the technical competence and democratic legitimacy of the siting agencies and contractors. This frame would prove to motivate significant active opposition in the candidate communities. The site selection processes and the generation of local opposition demonstrate the multifaceted nature of local power. In seeking to avoid opposition, siting authorities selected candidate sites in part on demographics thought to be associated with resources of political power, such as income. And in adding such considerations to the site selection criteria, the authorities were leveraging their power over the decisionmaking process, or "rules of the game." Yet opponents were able to leverage the power of widely shared meaning by painting these strategies as incompetent, illegitimate, and unjust.

Chapters 4 and 5 take a closer look at the mobilization of active local opposition to LLRW facilities across the 21 candidate counties. Even though opposition was prevalent in each of the candidate counties, the number and type of collective acts of public opposition they were able to muster varied. Newspaper analysis of collective opposition events reveals a wide range of frequency in events of active opposition across these cases, as well as significant variation in the predominant tactics employed to oppose LLRW site proposals. Interestingly, just as state siting authorities failed to identify acquiescent communities, theories of social movements have difficulty explaining variations in the frequency and type of active opposition across these cases. If the siting authorities overlooked the local power of meaning, social movement theory undersells the power of movement. By overemphasizing static, preexisting community characteristics and political opportunities, the classic social movement model misses the dynamic relationships between and among political actors that shape political outcomes. The identification and analysis of repeated patterns of social mechanisms across carefully paired candidate counties, based on scores of in-depth interviews, lend a detailed understanding of mobilization by different political actors in a variety of contexts.

Chapter 6 examines the effects the frequency and type of active local opposition had on implementation progress across the 21 candidate counties. Active local opposition by citizens and local governments effectively halted implementation of the LLRWPA, although not all candidate counties were removed from the siting process at the same rate or for the same reasons. Although none of the counties came to host an LLRW facility, some were dropped early in the implementation process, whereas others had actually been licensed for construction before implementation failed. There is a significant negative relationship between the number of collective acts of public opposition in a candidate county and progress toward implementation. The more active the opposition, the quicker the siting process ended for the candidate county. The number of disruptive acts of collective opposition was not, however, significantly related to implementation progress. Qualitative evidence demonstrates that early and active opposition of all types quickly drew support from local and then state officials, which ultimately halted the siting process.

The most dramatic example of this occurred in New York, where LLRW opponents eventually won the support of Governor Mario Cuomo, who successfully challenged the constitutionality of the LLRWPA and its amendments. In *New York v. United States* (1992), the U.S. Supreme Court struck down the most stringent incentive for states to create new LLRW disposal sites—the "take title" provision, which would have punished states noncompliant under the LLRWPA implementation milestones by requiring them to assume liability for all commercial wastes generated within their borders. In the absence of the "take title" incentive, many states facing active local opposition halted the LLRW siting process. At this point, states began competing to avoid the creation of an LLRW facility within their boundaries. Even states that had designed voluntary site selection approaches and offered generous compensation packages failed to find willing hosts. Eventually, states that had made significant implementation progress because they had not faced local active opposition began to seek ways to stall or halt the LLRW siting process. In these cases, state-level officials intervened to halt the siting process once implementation was failing nationwide, out of a fear that the first newly constructed LLRW site would ultimately be a national site in perpetuity.

Interestingly, in this context, willing hosts did emerge to dispose of the nation's commercially generated LLRW. They were not, however, new sites created under the LLRWPA. Chapter 7 concludes by describing the LLRW disposal situation that settled in after the states and regional compacts failed to build new sites. The two states hosting existing LLRW facilities continued to accept waste while exercising their authority to levy surcharges and regulate the flow of out-of-state waste. Washington accepted waste only from member states of the Northwest and Rocky Mountain Compacts, while South Carolina joined a compact with New Jersey and Connecticut but continued to accept waste from other states for a hefty fee until 2008. In 1995, Utah licensed a private LLRW facility outside of the LLRWPA but under agreement with the Northwest Compact at a repurposed site for hazardous and naturally occurring radioactive waste. This facility is operated by the EnergySolutions company and is licensed by Utah to accept only the least radioactive category of LLRW. The Northwest Compact and Utah have agreed to allow the facility to accept this category of waste from states outside of the compact. As this book goes to press, Texas seems poised to follow Utah's example and license a private LLRW disposal facility in an area already used for hazardous and radioactive waste processing. Texas is joined in a compact with Vermont and will likely accept all categories of LLRW that the South Carolina site had accepted, while collecting the hefty stream of revenue that site collected.

This is a far cry from the evenly distributed network of more than a dozen regional LLRW disposal sites Congress envisioned when passing the LLRWPA. However, it does provide enough disposal to enable continued generating activities in a way that reaps considerable revenue and maintains significant authority for host states. Because it provides this authority, the LLRWPA is still an important part of the current waste disposal situation. The law is still supported by the NGA and industry groups representing nuclear utilities and disposal professionals. The unusual equilibrium that has emerged contains a mix of satisfying and dissatisfying aspects for both antinuclear groups and the nuclear industry. The long-term stability of this equilibrium remains an open question, subject to a recent federal court decision weakening state and compact authority over private LLRW sites, alternative proposals for LLRW disposal provided by the federal government, and signs that nuclear power could be on the verge of a renaissance. Whatever shape new policy contexts affecting LLRW disposal take, they will be layered upon the already complicated implementation dynamics of the LLRWPA that trickled down from the federal government to states and regional compacts before meeting powerful local opposition and bubbling back up the intergovernmental strata.

CHAPTER 2

THE HALF-LIFE OF FEDERAL RESPONSIBILITY: THE DEVOLUTION OF LLRW DISPOSAL

In 1979, a truck carrying tools and parts contaminated with radiation from the infamous near core meltdown at the Three Mile Island nuclear power plant in Pennsylvania made its way toward a low-level radioactive waste (LLRW) disposal facility in Barnwell, South Carolina. The truck was turned back, however, by Heyward G. Shealy, South Carolina's secretary of health and environmental control. Shealy made the following proclamation: "We take a lot of [nuclear] waste down here, but we don't want to take all of it for the whole country. ... I don't think our citizens are interested in having to take all the dangerous stuff from Three Mile Island" (O'Toole and Peterson 1979, A1).

The Barnwell facility was one of three national disposal facilities for commercially generated LLRW in the United States. Each of these national facilities was operated by a state under agreement status with the federal government. In 1979, the governors of Washington and Nevada, the other two states hosting LLRW disposal facilities, joined South Carolina in restricting disposal access.

This embargo forced an obscure and poorly defined category of radioactive waste onto the national policy agenda in a way that revealed expanding problems associated with the management of nuclear technology, as well as tensions over equity and authority in the U.S. federal system of government.

Low-level radioactive waste was, and continues to be, a residual "catchall" category of radioactive waste defined by what it is not. LLRW is radioactive by-product *other than* spent nuclear fuel, reprocessed nuclear fuel waste, transuranic waste from plutonium weapons fabrication, and mill tailings from uranium mines. (Spent nuclear fuel and waste from reprocessed nuclear fuel together constitute *high*-level radioactive waste.) LLRW is heterogeneous according source, physical and chemical properties, and risk. Nuclear power plants produce a majority of LLRW, by both volume and radioactivity, with a diverse array of materials including filters and resins used during water purification and cooling, protective clothing, tools and metals, instrumentation, and wiring used near the nuclear fuel. Industry applications of nuclear technology ranging from the sterilization of contact lens solution to thickness measurements of paper and disposable diapers also produce LLRW. Hospitals and research institutions produce the smallest share of LLRW by volume and radioactivity from applications ranging from cancer treatment to carbon dating.

Despite its name, this strange category of waste is not necessarily characterized by either a low level of radioactivity or a low risk to human health. Shipment analysis at LLRW facilities has illustrated that the radioactive content within this waste stream varies widely (Parker 1988, 92–93). LLRW can be solid, liquid, or gaseous. It can be combustible or stable. Materials in LLRW can emit alpha, beta, or gamma radiation. Some LLRW decays in half-lives of more than 100,000 years, whereas other forms have half-lives of just 15 years. Some forms of LLRW are benign enough to handle, whereas others, such as the irradiated reactor core shroud and other metal parts inside a reactor, could be instantly fatal to a human without shielding.

Large-scale structural work at nuclear power plants, whether unplanned as in the case of the Three Mile Island accident or planned decommissioning, produces large quantities of LLRW necessitating disposal space. In 1979, the states hosting such space were no longer willing to take the waste. By 1980, the LLRW embargo moved Congress to identify the "availability of disposal capacity" for LLRW as a "national crisis" (U.S. House Committee 1980, 25).

Public policy scholars have long recognized that the declaration of a crisis or problem reflects more than an objective set of conditions. Problems are defined or socially constructed by political actors, who often are competing to promote different interpretations of and solutions to a set of conditions (Stone 1997; Baumgartner and Jones 1993). LLRW, and radioactive waste in general, was produced and dumped haphazardly in unlined trenches or the ocean for decades before the issue was defined as a problem in the policy arena. The 1979 LLRW embargo prompted the federal government to define the "crisis" as one of limited safe waste disposal capacity at existing sites that threatened the continued development of nuclear technologies. Representative Morris K. Udall, chair of the House Subcommittee on Energy and the Environment, conveyed this problem definition in a 1979 hearing on radioactive waste: "I submit that in the long run we will not have nuclear power as a major energy source here if the people are not sure that the nuclear system is reasonably safe. That confidence will never exist until we are satisfied we are able to manage our nuclear wastes" (U.S. House Subcommittee 1979a, 1).

The initial congressional solution to flow from this problem definition was to order the Department of Energy to expand LLRW disposal capacity by creating additional national LLRW disposal sites. State governments, however, contested this problem definition by adding two additional wrinkles to the perceived need for expanded disposal capacity: authority for and equity among states. The states, led by the LLRW facility hosts, argued for a more equitable distribution of disposal sites among them, over which individual states would have responsibility. Both the National Governors Association (NGA) and the National Conference of State Legislatures (NCSL) recommended that instead of creating more national LLRW disposal sites, Congress should cede responsibility for the disposal of LLRW to the states, requiring each state, either alone or in a regional compact, to provide LLRW disposal capacity. The states succeeded in persuading Congress to devolve responsibility for establishing new LLRW disposal facilities with passage of the Low-Level Radioactive Waste Policy Act (LLRWPA) during the last days of a lame-duck session in late December 1980. The act declared, "Each State is responsible for providing for the availability of capacity either within or outside the State for the disposal of low-level radioactive waste generated within its borders" (42 U.S.C. 2021d).

This was a dramatic policy shift. The LLRWPA stood in stark contrast to a nuclear policy characterized by many scholars as a federal "subgovernment" or "policy monopoly" dominated and jealously guarded by a small group of elites (Baumgartner and Jones 1993; Duffy 1997; Temples 1980). For decades, the federal Atomic Energy Commission (AEC) had successfully argued that states were not qualified to supervise LLRW disposal (Mazuzan and Walker 1984). As late as 1973, opponents of devolving disposal authority to the states, such as Representative Craig Hosmer from California, had successfully argued, "If we are going to give every Tom, Dick and Harry in every state legislature who wants to have his gummy fingers go over the licensing and regulation of atomic waste, then we are really going to get ourselves in a mess" (*Congressional Quarterly Almanac* 1973, 594).

What had changed in 1980? Why was the federal government willing to give the states responsibility over a significant aspect of nuclear policy? Why were the states asking to receive this responsibility? What were the implications of the problem definition written into the LLRWPA solution?

Answering these questions requires an understanding of the broader historical context of U.S. nuclear policy, with particular attention to the Atomic Energy Act and the politics that sprang from it. The Three Mile Island accident and ensuing LLRW embargo were merely proximate causes or even symptoms of more than three decades of politics associated with commercial uses of nuclear technology. Several authors have provided rich historical accounts of nuclear politics, while refining frameworks for understanding policy change (see Temples 1980; Baumgartner and Jones 1993; Duffy 1997). Baumgartner and Jones chronicled nuclear politics from the 1950s to the 1980s to establish their "agenda-setting model" of policy change from a highly centralized, consensual, closed policy arena promoting the expansion of nuclear power to an arena open and vulnerable to challenges from opponents. They presented the following sequence of events: "opponents exploited divisions within the community of experts; images in the popular media changed; opponents were able to obtain the attention first of regulators and then of Congress, the courts, and state regulators; finally, the market responded." Last of all, "public attitudes toward nuclear power responded to elite activity" (1993, 79).

But what precedes this sequence? How did the opposition materialize? Why were there divisions within the community of experts? Why did images in the popular media change? Why were other actors, such as Congress, the courts, and state regulators, ready to get involved?

Duffy (1997), the most thorough account of nuclear energy policy in the United States, provides a hint that is worth exploring further: Congress entrusted the AEC to both promote and regulate civilian nuclear power under the Atomic Energy Act. Analysis of policy types, as pioneered by Lowi (1985), would mark this law as setting the AEC at cross-purposes—conflating distributive and regulatory goals within the same agency. Distributive policies are often associated with patronage through the provision of subsidies, privileges, or facilities, while regulatory policies impose obligations and sanctions. Identifying policy types serves to remind us that policy makes politics.

Garrison Keillor once wrote "all of the Norwegians were Lutherans, of course, even the atheists—it was a Lutheran God they did not believe in" (1998, 17). The Atomic Energy Act, in its multiple manifestations, was like Lutheranism among Norwegians. It shaped not only the practicing adherents of the nuclear policy monopoly, but also its opponents. The policy shaped the political conflict that ultimately spelled the demise of the AEC and led to the rising involvement of states in nuclear issues. In Hogwood and Peters's list of pathologies of public policy (1985), the Atomic Energy Act would qualify as a "congenital disease" of the type "crippled at birth." Internal conflict among technical experts, physical disasters with nuclear energy, and plummeting public support for nuclear plants all grew out of seeds planted in a dysfunctional nuclear policy that conflated distributive and regulatory missions.

A DISTRIBUTIVE POLICY MONOPOLY: THE BIRTH OF NUCLEAR ENERGY POLICY, 1942–1945

A 12-foot Henry Moore sculpture sits on the east side of Ellis Avenue between 56th and 57th Streets in Chicago. The massive bronze sculpture, titled *Nuclear Energy*, marks the former location of a University of Chicago squash court.

On December 2, 1942, this unlikely space below the bleachers of Stagg Field hosted the world's first controlled nuclear chain reaction. At the very least, some squash playing regulars must have noticed Enrico Fermi's nuclear "pile" of graphite and uranium on their court. And three years later, ranchers in Alamogordo, New Mexico, must have noticed something unusual at 5:30 a.m., when the world's first atomic bomb split the nuclei of a baseball-size sphere of plutonium. The bomb turned the asphalt around the blast site into a greenish glass and sent a mushroom cloud up 41,000 feet in the air.

Still, all accounts of the U.S. quest for an atomic weapon emphasize the secrecy of the so-called Manhattan Project. "Secret cities" of scientists sprang up in Los Alamos, New Mexico; Oak Ridge, Tennessee; and Hanford, Washington to develop an atomic weapon. Major General Leslie R. Groves, the army engineer in command of the project, reported directly to the secretary of war and the president's chief of staff, leaving most of the military establishment ignorant of this important new technology.

Early nuclear policy not only was secretive and highly technical, but it also was distributive. Archetypal distributive policy areas are public works projects such as the construction of roads, bridges, railroads, and canals, and the Manhattan Project was a highly technical and focused public works project. The government contracted with private industry to develop an atomic weapon. The DuPont Corporation constructed and operated the Hanford plutonium production plant. Monsanto Chemical Corporation and a Union Carbide subsidiary operated other production facilities in Tennessee. Numerous other corporations across the country contributed to the top-secret project and gained significant economic benefits from the government-business partnership (Burns and Briner 1988; Duffy 1997). The project established an ideal policy monopoly, typical of distributive policy areas. It conferred specific, unconditional benefits to a small group. The costs were diffuse and the project was secret. More important, the project had no rules or regulatory aspects, because of the urgency of war.

In 1945, the Manhattan Project produced "little boy" and "fat man," the atomic weapons that U.S. B-29 bombers dropped on Hiroshima and Nagasaki in the early days of August. World War II ended less than a month later, and the nuclear policy monopoly disbanded. However, nearly a year after the end of World War II, Congress passed the Atomic Energy Act of 1946 (also known as the McMahon Act) to continue nuclear weapons development. This act created an entrenched policy monopoly every bit as strong as the secret wartime arrangement, and it expanded the scope of research and development beyond weapons to nuclear energy production.

A BIFURCATED POLICY MISSION: DISTRIBUTIVE AND REGULATORY GOALS IN U.S. NUCLEAR ENERGY POLICY, 1946–1960

The extension of U.S. nuclear policy from weapons production into energy production complicated the previous policy mission of militarization. The Atomic Energy Act of 1946 was not simply distributive in nature, but it was also regulatory. The McMahon Act was "an act for the development and control of atomic energy." The "development" included distributive efforts in research and development of nuclear energy. Yet the "control" of atomic energy included regulatory efforts to "protect health, to minimize danger from explosion and other hazards to life or property," and "to require the reporting and permit the inspection of work performed." Not only were these functions of development and control pronounced in the same statute, but they also were charged to a single division (Research) within a single agency (the AEC). This bifurcated policy mission was to have lasting consequences for the next 30 years.

The organic Atomic Energy Act of 1946 had the seeds of both distributive and regulatory policy, but the distributive seeds flourished and the regulatory seeds remained dormant. This is not surprising, given the fact that potential uses for nuclear energy were still unknown. The act states that "the uses of atomic energy for civilian purposes ... cannot now be determined" (79 P.L. 585). Until the distributive tasks of research, development, and promotion actually produced viable civilian uses for nuclear power, there was little to regulate except the research itself.

Flourishing Distributive Mission

The Atomic Energy Act of 1946 sowed the seeds for the involvement of private industry in the development of commercial nuclear power generation. The act stated that the "utilization of atomic energy" should be directed toward, among other things, "strengthening free competition in private enterprise" (79 P.L. 585). The 1946 act gave the newly created AEC ownership of all nuclear research laboratories and other atomic resources, and exclusive control of all information pertaining to nuclear technology.

Amendments to the act in 1954, under the Eisenhower administration, encouraged the AEC to utilize these atomic resources in coordination with private industry. Eisenhower saw the civilian production of nuclear power as an "atomic Marshall Plan" that could help win the Cold War. He argued that private industry could develop nuclear power faster and cheaper than the federal bureaucracy (Eckstein 1997). In his famous Atoms for Peace speech before the United Nations in December 1953, Eisenhower explained that the U.S. monopoly on nuclear power had ended, the USSR was devoting extensive resources to nuclear weapons, and other nations would soon develop nuclear weapons as well. He said that the United States wanted to move the world "out of the dark chambers of horror and into the light" and adapt nuclear technology "to the arts of peace ... into a great boon for the benefit of all mankind" (Eisenhower 1953). One month later, in his Budget Message, President Eisenhower asked that the Atomic Energy Act of 1946 be amended to "encourage wider participation by private industry in atomic development for peaceful purposes," arguing that the development of nuclear power required private investment "to assure the greatest efficiency and progress" for the least public cost (*Congressional Quarterly Almanac* 1954, 536).

In 1953, Lewis Strauss, a Wall Street financier, became President Eisenhower's special assistant for atomic energy and the chair of the AEC. The agency submitted a bill to "promote and encourage free competition and private investment" in nuclear power generation. A group of 75 electrical utility executives proclaimed that "it takes private industry to do a big job like atomic development." After a lengthy debate in the Senate and two conference committees, the resulting Atomic Energy Act of 1954 was a victory for Eisenhower and the utility companies, which would benefit directly from government-subsidized nuclear research and development. Eisenhower proclaimed that it was "time to draw more specifically into the national atomic energy program the initiative and resources of private industry," and that this legislation would "relieve the taxpayer of the enormous cost of the commercial aspects of the enterprise" (*Congressional Quarterly Almanac* 1954, 535, 546).

The AEC exercised its distributive largesse by funding private research initiatives on nuclear power, sometimes channeling cash to private companies for this purpose. The agency subsidized uranium enrichment and created several prototype reactors to demonstrate the technological viability of nuclear power (Eckstein 1997). It provided companies with enormous subsidies to ensure plant construction. For example, Westinghouse constructed the Shippingport, Pennsylvania, reactor at a cost of $55 million. The Duquesne Light Company paid $5 million for the new plant, and the AEC paid the remaining $50 million (Eckstein 1997). The agency also provided free uranium fuel for up to seven years and guaranteed the purchase of all plutonium produced at a fixed price. Several of the first commercial plants were built by General Electric and Westinghouse under this program, including Indian Point-1 in New York, Fermi-1 in Ohio, and Dresden-1 in Illinois (Carter 1988).

In 1957, the Price-Anderson Act amended the Atomic Energy Act of 1954 and removed the final economic obstacle to private investment in nuclear power generation. The act allowed U.S. funds to cover "a portion of the damages suffered by the public from nuclear incidents" and enabled Congress to "limit the liability of those persons liable for such losses" (85 P.L. 256). Between 1954 and 1959, utilities ordered 15 nuclear reactors. By 1962, the federal government had spent more than $1.2 billion to encourage reactor development—double the amount of private investment. Years later, the Department of Energy estimated that without subsidies, nuclear power would be at least 50 percent more expensive (Duffy 1997).

Foundering Regulatory Mission

The distributive characteristics of the Atomic Energy Act ultimately overwhelmed the regulatory tasks laid out in the law, but the regulatory aspects of the AEC mission were still codified. The Atomic Energy Act charged the AEC, vaguely, to develop licensing procedures and penalties for the mishandling of nuclear materials. The law accounted for nuclear waste for the first time, by defining by-product materials as "any radioactive material yielded in or made radioactive by exposure to the radiation incident to the processes of producing or utilizing fissionable material." By-product material was to be regulated with "such safety standards to protect health as may be established by the Commission" (79 P.L. 585). The law explicitly stated that the AEC was subject to the Administrative Procedure Act, with its commensurate hearings and public notice and consent provisions.

Some members of the nuclear establishment attempted to raise regulatory concerns to counter the distributive trend favoring private industry during the Eisenhower administration. Several scientists intimately involved with nuclear research doubted that the push for civilian production of nuclear power would satisfy the AEC's regulatory responsibility to provide for public safety. Former AEC chair David Lilienthal asked, "Should not a program of large-scale atomic reactors wait at least until it has been demonstrated that this waste problem has been 'licked'?" He continued, "So long as the safe handling of radioactive waste for a large-scale national atomic power program is still in the research stage ... the future of atomic energy as a major reliance for civilian electricity is in grave doubt" (Carter 1988, 80).[1] As late as 1952, AEC chairman Gordon Dean stated that the time was not right for opening atomic development to private industry. He said, "The need for restraints on private industry in the atomic program is greater than the need for doing away with them" (*Congressional Quarterly Almanac* 1954, 535). Robert Oppenheimer, the former chair of the General Advisory Council overseeing the AEC, reportedly had his skeptical comments on nuclear power censored (Eckstein 1997, 43).[2] James Conant, a chemist who served as chair of the National Defense Research Committee overseeing the Manhattan Project in the 1940s, also opposed nuclear power. He wrote that experimental reactors produced waste that "had presented gigantic problems—problems to be lived with for generations" and that a "self-denying ordinance [is] but common sense" (Carter 1988, 53). This opposition was exiled or silenced as Lewis Strauss

---

[1] Lilienthal later lent his support to opponents of a 1962 plan to build a nuclear reactor in Queens, New York.

[2] Oppenheimer's loyalty to America was ultimately questioned during the height of McCarthyism, and his security clearance was revoked in 1954.

promoted the distributive aspect of the Atomic Energy Act at the expense of the regulatory aspect.[3]

Although the Atomic Energy Act of 1954 is best known for opening nuclear technology to private enterprise by allowing corporations to attain licenses for nuclear power production and patents over nuclear technology, it also laid out a much more explicit regulatory role for the AEC. To be sure, the act established a clearly distributive purpose: "to encourage widespread participation in the development and utilization of atomic energy for peaceful purposes." But this purpose was immediately qualified to be within the "maximum extent consistent with the common defense and security and with the health and safety of the public." In the opening lines of the act, before any new mention of private enterprise, the 1954 act stated that "the processing and utilization" of fissionable materials "must be regulated in the national interest ... to protect the health and safety of the public" (42 U.S.C. 1801). In fact, half of the findings in the opening lines of the act are regulatory in nature. More important, the act details licensing procedures, penalties, and administrative procedures that charge the AEC with regulatory responsibilities.

Yet the AEC neglected its regulatory mission. Eckstein argued that the most important AEC action facilitating commercial nuclear reactors was "rubber-stamp licensing and regulation of atomic reactors" (1997, 33). As one AEC attorney said, "Nobody really even thought safety was a problem. They assumed that if you just wrote the requirement that it be done properly, it would be done properly" (Ford 1982, 42). The congressional Joint Committee on Atomic Energy (JCAE) repeatedly pressured the AEC to ease the licensing process (Carter 1988; Duffy 1997; Mazuzan and Walker 1984). Eisenhower's AEC chair, Lewis Strauss, stated that regulations on nuclear power "should not impose unnecessary limitations or restrictions upon private participation" and that the "AEC's objective in the formation of regulations was to minimize government control of competitive enterprise." As of 1960, AEC commissioners estimated that only one-sixth to one-third of their time was devoted to regulatory matters, and the AEC director of regulation said that his staff was

---

[3] The experts dissatisfied with the AEC lack of attention to safety did not, however, disappear. Lilienthal and Conant would both lend legitimacy and support to nuclear power opposition in the 1960s through essays and speeches.

small enough to "meet in a phone booth" (Duffy 1997, 37, 43–44). Carroll Wilson, the AEC's first general manager, recalled that "nobody got brownie points for caring about nuclear waste" (McCutcheon 2002, 8).

REGULATORY SHORTCOMINGS: ACCIDENTS, INCIDENTS, AND THE DETERIORATING NUCLEAR POLICY MONOPOLY, 1961–1979

The robust distributive efforts of the AEC successfully created a booming commercial nuclear power industry in the 1960s. However, the lack of matching regulatory muscle from the AEC during this period set the stage for nuclear accidents, mismanagement, and internal dissension among nuclear experts. Events such as these diminished public trust in nuclear power and moved states and other governmental actors to encroach on the once solid nuclear policy monopoly of the AEC.

The China Syndrome and the Decline of Nuclear Power Plant Orders

In the 1979 film *The China Syndrome,* Jack Lemmon, playing a nuclear power plant manager, coyly delivered the following barroom pickup line to a reporter played by Jane Fonda: "When you turn on your lights, think 10 percent me." By 1979, nuclear power was an established player in the electrical utility business, with 69 fully operating generators. The film depicted an arrogant disregard for public safety on the part of utility companies and government regulators. This lack of regulatory concern led to a near meltdown at a California nuclear plant. The "China syndrome," which would occur if the nuclear core of a power plant melted down into the ground until it hit groundwater and exploded back through the ground as radioactive steam, was narrowly averted. As Lemmon's character explained in the film, the China syndrome "renders an area the size of Pennsylvania permanently uninhabitable."

This film was released just two weeks before the accident at the Three Mile Island nuclear plant near Harrisburg, Pennsylvania. In this real-life flirtation with the China syndrome, the experienced staff at the Metropolitan Edison Company violated Nuclear Regulatory Commission (NRC) regulations when they shut down auxiliary cooling pumps for maintenance two weeks prior to the accident. An investigative commission later found that the power plant staff "did not have sufficient knowledge, expertise and personnel to operate the plant or maintain it adequately." A pressure indicator stuck, giving the control room erroneous readings. The reactor containment vessel failed to isolate radioactive water, which was improperly vented as steam. Finally, a hydrogen bubble formed on top of the reactor as the temperature of the water rose so high that the water molecules were split into oxygen and hydrogen. The bubble posed a risk of explosion. Eventually the bubble shrank, although scientists could not explain why, but the danger of a meltdown had been averted (*Congressional Quarterly Almanac* 1979). Thousands of people evacuated the area, and Walter Cronkite told the nation that "the world has never known a day like today. . . . [T]he specter was raised that perhaps the most serious kind of nuclear catastrophe, a massive release of radioactivity," could occur (Stephens 1980, 4).

The Three Mile Island accident was really the last and most dramatic event in a parade of embarrassing nuclear blunders prior to passage of the LLRWPA. The AEC preference for distributive promotion over regulation led to a proliferation of nuclear power plants. However, problems at these plants and with the waste they generated ultimately led to internal dissension in the scientific community and public relations disasters for the nuclear industry and federal bureaucracy.

By 1979, the press coverage of nuclear power was predominantly negative, as were congressional hearings (Baumgartner and Jones 1993; Weart 1988), and a powerful antinuclear movement had taken hold both in the United States and abroad (Eckstein 1997; Dalton et al. 1999; Lesbirel 1998). The stock market value of nuclear utilities had fallen, there were no new orders for nuclear power plants, and many orders placed in previous years were canceled (EIA 2001). Figure 2.1 shows the trend that all scholars of nuclear politics confront: the rise and fall of orders for commercial nuclear power plants.

Private investment in nuclear power plants had increased dramatically in the 1960s and reached a peak in the early 1970s. Utilities ordered 136 new nuclear plants between 1971 and 1976. However, the number of new orders fell drastically in the late 1970s and evaporated entirely in the 1980s. The utilities also canceled many of the orders they had placed in earlier years. A pattern of unheeded regulatory warnings preceded this dramatic decline of nuclear power.

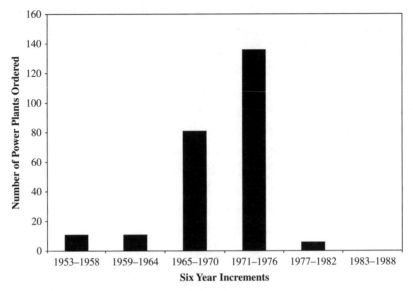

Figure 2-1. Orders for Nuclear Power Plants, 1953 – 1988
Source: EIA (2001, 257)

Unheeded Regulatory Warnings

Several unheeded regulatory warnings stand out in the 20 years preceding the decline of nuclear power in the United States. Perhaps the earliest unheeded warning came from a toothless advisory committee to the AEC that was established in a 1957 amendment to the Atomic Energy Act. The Advisory Committee on Reactor Safeguards (ACRS) was staffed by part-time officials paid on a per diem basis. The ACRS was charged to make recommendations on the health and safety concerns surrounding new nuclear power plants. In an early regulatory effort, the ACRS refused to give its approval to the Fermi-1 reactor on Lake Erie, 30 miles from Detroit. The AEC ignored ACRS safety concerns and attempted to suppress the ACRS report. The Fermi-1 plant suffered a partial core meltdown just months after it opened in 1966.

The ACRS was not the only group of experts issuing safety warnings to the AEC. The National Academy of Sciences (NAS) convened a committee on nuclear waste disposal during this same time period. The committee argued that "unlike the disposal of any other type of waste, the hazard related to radioactive waste is so great that no element of doubt should be allowed to exist regarding safety. Stringent rules must be set up and a system of inspection and monitoring instituted. Safe disposal means that the waste shall not come in contact with any living thing" (Carter 1988, 55).

The committee claimed that current nuclear waste disposal practices were inadequate. Six years later, Harry Hess, a member of the NAS committee on nuclear waste, sent an unsolicited letter to the AEC to express the opinion that "no existing AEC installation" generating nuclear waste had "a satisfactory geological location for the safe local disposal of such waste," and the present disposal methods for dealing with nuclear waste were unsafe. Hess argued that "approved plans for the safe disposal of radioactive wastes be made a prerequisite for the approval of the site of any future installation of the AEC or under AEC jurisdiction" (Carter 1988, 55).

These warnings proved prescient in 1959. In that year, the JCAE held a hearing on waste disposal in which Herbert M. Parker, the manager of the Hanford facility, testified that the current practice of storing high-level waste in tanks should be continued indefinitely. He argued that when the tanks wear out after 50 years, the contents could be pumped into a new set of tanks. Although Parker stated that no tank had ever leaked, it was soon discovered that 35,000 gallons of waste had leaked six months before his testimony. The Hanford tanks continued to leak thousands of gallons over the next two decades. One year after the Hanford leaks were publicly identified, a quarry in Weldon Spring, Missouri, that held nuclear waste from the first controlled chain reaction was contaminating groundwater. Public revelation of these leaks provided the first indication that the AEC had not adequately regulated the disposal end of the nuclear fuel cycle.

The AEC's disregard for regulatory responsibility also caused frustration among the scientists who worked under the Atomic Energy Act's regulatory provisions. In one well-known example, two scientists at the AEC's Lawrence Livermore Laboratory were asked to determine whether there was a correlation between nuclear fallout from atmospheric testing and infant mortality rates. AEC scientists John Gofman and Arthur Tamplin found that there was a significant correlation and that the AEC standards were inadequate. Instead of accepting the results, however, the AEC sought to discredit the scientists (Gofman and Tamplin 1971). In another example, AEC researchers who raised safety questions about the emergency cooling systems on nuclear power plants, as did Philip Rittenhouse, found that their reports were ignored and their projects canceled, and they were assigned to different tasks (Duffy 1997).

Scientists who were frustrated that their warnings went unheeded eventually took part in public hearings to oppose nuclear power plants. In 1971, several AEC staff members at Oak Ridge National Laboratory, disgruntled at the lack of agency attention to health and safety regulations, leaked experimental results that demonstrated that AEC regulations for emergency core cooling systems were inadequate.

One experiment found that fuel rods did not respond as expected during a loss-of-cooling accident. When scientists testing the safety of the cooling system reported the results, the AEC canceled the rest of the research and reassigned the scientists to different projects. More than 30 scientists and engineers, many from the AEC regulatory staff, publicly revealed their doubts concerning the safety of the emergency core cooling systems. This controversy was covered by the major television networks as well as reputable technical journals. When the AEC held public hearings on the issue in an attempt to discredit the opposition and defuse the issue, greater controversy emerged. The AEC lost a great deal of credibility when the hearings revealed that the agency subcontracted work on health and safety systems to nuclear power vendors—the very industry the AEC was charged to regulate (Duffy 1997).

In 1975, 2,300 scientists signed a petition urging the United States to reduce the construction of nuclear power plants. Some of the more notable signatories were James B. Conant, George B. Kistiakowsky, and Victor Weisskopf, all of whom worked on the Manhattan Project (Duffy 1997). In that same year, three longtime managing engineers from the General Electric nuclear reactor division resigned over their concerns for public health and safety. In 1976, two NRC engineers resigned, alleging that the agency covered up problems "of far-reaching significance" at nuclear power plants (*Congressional Quarterly Almanac* 1976, 92).

The Rise of Regulation

Technical failures in the wake of unheeded warnings led Congress to create an ever-increasing number of regulatory statutes governing nuclear power, which brought new governmental actors into the realm of nuclear policy. In 1959,

Congress established the Federal Radiation Council (FRC) to involve other bureaucratic agencies and the NAS in nuclear policy oversight. This enabled the Fish and Wildlife Service (FWS) to pressure the AEC to address the thermal pollution caused by nuclear plant releases of heated water into lakes and rivers. The AEC repeatedly rebuffed the warnings of the FWS, arguing that it had no regulatory authority over the nonradiological aspects of nuclear power (Duffy 1997). After years of struggle, the AEC was finally bound to consider all of the environmental impacts of nuclear power plants in 1971 when the D.C. Circuit Court of Appeals found that the AEC's licensing process and decisionmaking procedures violated the National Environmental Policy Act (NEPA) (*Calvert Cliffs' Coordinating Committee, Inc. v. U.S. Atomic Energy Commission* 1971).

This decision opened the AEC to the NEPA procedures for environmental regulation and public hearings. This was the most significant change in a growing list of legislation that affected AEC jurisdiction over nuclear policy. Duffy (1997) lists eight laws created between 1966 and 1977 that impinged on AEC dominance in the nuclear arena. For example, the Water Quality Improvement Act of 1970 gave the Environmental Protection Agency (EPA) authority to regulate the water discharged from nuclear power plants. Seven years later, in the Clean Air Act Amendments, EPA was given authority over transient radiation emitted through the air. Other laws gave the FWS and the Federal Emergency Management Administration (FEMA) jurisdiction over some areas of nuclear power (Duffy 1997). During these years, Congress also began adding money over and above the AEC requests for regulatory activities and cutting money requested for nuclear power plant construction activities. These changes were occurring during the creation of a new era of environmental policy, with more than 20 new laws expanding federal regulatory authority over environmental issues ranging from pollution control to endangered species protection.

The States Seek a Regulatory Role

While Congress paved the way for national governmental actors such as the FWS and FEMA to get involved in regulating nuclear energy policy, state governments were clamoring for a regulatory role as well. Congress did formally recognize the need for cooperation and coordination between the AEC and state regulatory agencies concerning radiation hazards and nuclear waste in a 1959 amendment to the Atomic Energy Act. However, this act did little to open the AEC to state regulatory agencies. For example, when Kentucky applied to the AEC for regulatory authority over radioactive waste disposal in 1962, the AEC denied the state's application. This action drew a strong condemnation from the Southern States Governors Conference (Burns and Briner 1988).

The lack of serious regulatory efforts by the AEC intensely frustrated state governments—particularly in the area of nuclear waste. The AEC ignored NAS warnings about a plan to bury high-level radioactive waste in caverns near the Savannah River Plant in South Carolina. The caverns were a mere 500 feet from the Tuscaloosa Aquifer, a major source of water for the entire southeastern United States. Georgia governor Jimmy Carter led a concerted effort of Georgia and South Carolina officials against the project. Speaking for the southern governors, Carter said that the AEC was "exhibiting a dangerously indifferent" and "irresponsible" attitude (Carter 1988, 61).

A similar issue during the late 1960s involved the Kansas state government. After a catastrophic fire at the Rocky Flats plutonium facility near Denver (yet another embarrassing accident for the AEC), the AEC tried to dispose of this debris at the National Reactor Testing Station (NRTS) near Idaho Falls.[4] However, the Idaho governor and two U.S. senators representing Idaho protested because they were promised that all of the waste at the NRTS would be taken to a permanent salt-mine disposal facility. In this context, the AEC announced that a salt mine in Lyons, Kansas, would be converted into a permanent repository. But the AEC had not adequately examined the geology and hydrology of the area surrounding the site.

Geologists working for Kansas soon released news that a salt mine one-third of a mile from the AEC site was leaking water. An unknown number of oil and gas well holes had penetrated the salt mine. The salt mine itself was mined in such a way that no pillars remained to support the cavern. The water leak revealed an imminent risk that the whole site would collapse into a huge sinkhole and flood with water. The failure of the AEC to discover this weakness

---

[4] In 1969, scrap from plutonium metal used to make nuclear weapons spontaneously ignited at the Rocky Flats plutonium plant. The resulting fire left more than 9,300 cubic meters of radioactive material.

in the site before it announced plans for a repository seriously injured the agency's credibility (Carter 1988).

When the AEC did involve states in nuclear waste disposal, the results were frustrating for those states. The AEC shipped spent fuel from Hanford to the West Valley Demonstration Project 30 miles south of Buffalo, New York, for reprocessing. West Valley was constructed and operated in 1963 by Nuclear Fuel Services (NFS), a newly formed private corporation. The AEC devolved responsibility for the project to the state of New York, but it maintained control of the price NFS could charge for its services. The AEC kept the price of fuel reprocessing so low, to the delight of waste producers, that NFS did not have enough capital to use the corrosion-resistant stainless steel tanks that British and French reprocessing facilities employed to store high-level waste. The NFS tanks failed to contain the waste, creating cleanup problems that are still ongoing today (Carter 1988).

Another site at Maxey Flats, Kentucky, worked under state "agreement" status with the AEC. The agency approved the burial of transuranic waste at this site, even though the NAS and AEC's own scientists had already recommended that this waste should be placed in a deep geologic repository. The state of Kentucky, not the AEC, finally halted acceptance of this waste when evaporated water from the burial trenches contaminated the local milk supply (Carter 1988).

Many states took aggressive action to confront the AEC. Most notably, California adopted legislation that placed a moratorium on nuclear power plant construction until safe nuclear waste disposal methods could be demonstrated. Minnesota issued regulatory standards for nuclear emissions that were much more stringent than those of the AEC. Vermont, New Hampshire, and Massachusetts challenged the AEC to meet state water quality standards.

In the early 1970s, the congressional debate on AEC authorization offered an opportunity for critics to attack the agency. In 1973, Representative Bertram L. Podell (D-NY) proposed an amendment to the AEC authorization that would have allowed states to set their own standards regulating the discharge or disposal of radioactive waste if they were more stringent than AEC regulations. In support, Representative Donald M. Fraser (D-MN) argued that states "are occasionally even wiser than some federal agencies—particularly an agency whose record leaves something to be desired, if I may say so" (Carter 1988).

The amendment was defeated after vigorous debate. In 1974, Senator Richard S. Schweiker (R-PA) unsuccessfully attempted to permit states to add additional insurance requirements for nuclear power plants.

In 1975, Congress abolished the AEC under the Energy Reorganization Act. The Nuclear Regulatory Commission (NRC) assumed the AEC regulatory responsibilities, while the Energy Research and Development Administration (ERDA) absorbed the distributive functions. Nearly 30 years after the contradictory Atomic Energy Act was first drafted, the regulatory and distributive aspects of nuclear policy were separated.

The change in federal agencies did little to alter the states' distrust of federal nuclear regulation. In 1975, ERDA began searching in earnest for a permanent high-level radioactive waste disposal site in the underground salt beds located in numerous states. ERDA began its search in Michigan, where it first stated that no disposal site would be constructed "if the people don't want them [nuclear wastes] there." Two weeks later, ERDA's general counsel reversed this statement and claimed that the agency was "reserving the right" to locate a disposal site despite state objections. This ignited state and local anger directed at ERDA (Carter 1988). Michigan congressman Philip R. Ruppe called the siting process a "bureaucratic sneak play," and the state's governor, William G. Milliken, successfully championed state legislation that forbade the disposal of nuclear waste in Michigan (U.S. House Subcommittee 1977). ERDA plans for repositories in Ohio, New York, Utah, Texas, Mississippi, Louisiana, Washington, Nevada, and New Mexico all met with vigorous state opposition. In several cases, ERDA made statements that offered state governments the final say on repository construction, only to reverse these statements at a later date. This generated a great deal of distrust at the state level and inspired state legislation that placed a moratorium on site exploration, forbade nuclear waste disposal or transportation altogether, or made the regulatory process an impossible labyrinth (Carter 1988; Greenwood 1982).

The immediate events that precipitated passage of the Low-Level Radioactive Waste Policy Act were placed in this larger context of tense relations between state governments and the federal bureaucracy responsible for nuclear waste. A history of regulatory neglect by the AEC and the resulting accidents and mismanagement concerning nuclear policy left the states clamoring for a greater regulatory role.

The States Ask for Regulatory Responsibility: The LLRW Policy Act

The OPEC oil embargoes sparked an energy crisis in the 1970s, which presidents Nixon, Ford, and Carter had to confront. Each of these presidents sought to revitalize nuclear power generation, yet creating a permanent disposal solution for nuclear waste remained an obstacle to this goal. President Carter promised to "work towards a policy for safe, permanent disposal of nuclear wastes" in his 1978 State of the Union Address (Carter 1978). As a governor, Carter had confronted the AEC over waste disposal issues. He understood the distrust of the states for federal authority over nuclear policy, whether it came from the AEC, NRC, ERDA, or the Department of Energy (DOE), which Carter created in 1977. For this reason, the administration did not entrust work on the new nuclear waste management initiative to the DOE. As one observer close to the administration noted, "This was in large measure the result of a perception that DOE's credibility—as the successor to the AEC and ERDA—was so low in this area and its ability to escape from past history was so constrained that it was simply not capable on its own of adequately reconstituting the nuclear waste management program" (Greenwood 1982, 20).

Instead, Carter created the Interagency Review Group (IRG) to formulate a nuclear waste management plan. The IRG consisted of officials from 13 executive branch departments and agencies. The group published its final report on radioactive waste disposal in the spring of 1979, just weeks before the accident at Three Mile Island intensified the political urgency of a nuclear waste solution. Once again, the federal bureaucracy responsible for nuclear power suffered a serious public relations disaster. The Kemeny Commission (1979) argued that "the NRC is not necessarily a mismanaged agency—it is an unmanaged agency."

In the wake of the Three Mile Island accident, many members of Congress shared Senator Gary Hart's sentiment that nuclear policy was a "continual encroachment on vital state prerogatives" (*Congressional Quarterly Almanac* 1979, 694). The Senate passed amendments to the NRC authorization bill that required the agency to notify state authorities when it transported nuclear waste across state lines. The final report of the IRG sought the creation of mechanisms and funds that enabled states to be informed participants in review and licensing proceedings for nuclear waste management (*Congressional Quarterly Almanac* 1979).

The national political climate was highly critical of federal efforts to regulate nuclear energy and radioactive waste, and sympathetic to state demands for more responsibility over disposal. This sentiment was not limited to radioactive waste, but was part of a growing call for intergovernmental reform that would devolve greater authority to the states. Some scholars of the time (e.g., Lowi 1978) wondered whether the increasing size and scope of the federal government had caused the United States to become a unitary rather than federal system. The bipartisan U.S. Advisory Commission on Intergovernmental Relations reported that the U.S. federal system was "endangered" by the "coercive character" of the federal government and the erosion of state and local authority (1981, 5). In the 1980 presidential election, Ronald Reagan made devolution a cornerstone of his governing philosophy.

In this context, Heyward G. Shealy of the South Carolina Department of Health and Environment Control ordered a convoy of trucks carrying radioactive waste from the Three Mile Island accident to turn back toward Pennsylvania. Richard W. Riley, the newly elected governor of South Carolina, explained why his state refused to accept the waste in an op-ed article in the *Washington Post*: "I believe it would be irresponsible for us to make it easy, by our acquiescence, for the nation to delay these decisions [on nuclear waste disposal]. South Carolina can no longer be the path of least resistance in seeking the national answer to nuclear waste disposal" (Riley 1979).

Low-Level Radioactive Waste: A Heterogeneous Residual Category

The filters, tools, equipment, dials, and gauges that had been contaminated by the Three Mile Island accident en route to South Carolina were not yet statutorily defined as "low-level radioactive waste." The Atomic Energy Act and subsequent amendments simply identified any radioactive material "yielded in or made radioactive by exposure to the radiation incident" as by-product material (42 U.S.C. 2011, Ch. 2, Sec. 11, e). Unlike the International Atomic Energy Agency (IAEA), which defines and classifies radioactive waste categories according to the radiation level and half-life, U.S. law classifies radioactive waste by source: mill tailings from extraction, spent fuel from reactors, transuranic materials from weapons production, and waste from fuel reprocessing. By-product material did not fit into any of these categories.

This system of definitions was rooted in the bifurcated nature of the AEC and the predominance of distributive over regulatory responsibilities. As one contemporary NRC employee, Mel Knap, told me in a 2004 interview, "Historically, the AEC was not concerned with the risk that nuclear waste posed to public health or the environment; it was concerned with developing nuclear power generation. So the definitions of waste are not based on risk; they are based on the process by which the waste is generated."

The LLRW waste shipment that was leaving the Three Mile Island plant for Barnwell, South Carolina, like all LLRW shipments in the United States, was heterogeneous in the risk it posed to human health. The radiation level of some of this material was just above the background levels found in nature, while other material, such as components from inside the reactor vessel, was highly radioactive. Historically, this broad category of waste was all treated with the same minimal standards. In the 1940s and 1950s, much of this waste was buried uncontainerized in shallow earthen trenches or dumped at sea. The rise of commercially generated LLRW from nuclear power generation led to the creation of six national disposal sites operated by states under agreement status with the AEC. In 1971, there were six sites receiving commercially generated LLRW. By 1979, only three disposal sites were still accepting waste: the Barnwell facility in South Carolina: a site in Beatty, Nevada: and the Hanford site in the state of Washington. Three other sites—in New York, Illinois, and Kentucky—were open just 13, 10, and 15 years, respectively, before water infiltration and container corrosion enabled radiation to escape containment.[5]

An LLRW Embargo

Three months after South Carolina refused to accept the Three Mile Island waste, Nevada governor Robert List closed the Beatty site to all LLRW. His action was prompted by three troubling incidents at the site. First, a truck carrying radioactive medical waste caught on fire at the site entrance and exposed 10 people to radiation. Second, another truck that was purported to be carrying dehydrated waste from a Michigan power plant arrived at the site

---

[5] For a more recent assessment of LLRW waste stream composition, risks, costs, and disposal practices in the United States, see National Research Council (2003).

leaking contaminated liquids. Finally, drums containing radioactive waste were found buried outside the site fence line. The closure of this Nevada site left just two remaining LLRW disposal sites open.

In October 1979, Dixie Lee Ray, the governor of Washington and former chair of the AEC, temporarily closed the Hanford LLRW facility. She said her decision was in response to a series of transportation and packaging incidents, including one shipment that was leaking radioactive cobalt. Ray successfully pushed a statewide referendum that ordered the site closed to all nonmedical out-of-state waste (Condon 1990; U.S. House Committee 1985). (The courts struck down this referendum later on constitutional grounds.)

Governor Riley of South Carolina soon began restricting the volume of shipments at the Barnwell facility and refused to accept any liquid waste. Riley argued that a six-month inspection of incoming LLRW showed that shipments were consistently above the allowable radiation limits and packaged improperly. He maintained that "these discrepancies were not being dealt with appropriately by federal authorities. ... A regulatory process that is not enforced is nothing more than sham" (1982, viii). Ralph R. DiSibio, the director of Nevada's Human Resources Department, said that the three host states had no alternative but to restrict access, noting that "the attention span of the people involved lasts as long as the closure" (*Business Week* 1979).

The LLRW facility closures gained a considerable amount of attention. Alfred P. Wolf, a senior chemist at Brookhaven National Laboratory, told *Business Week* that "whole areas of U.S. business could go down the drain" as a result. Jason M. Salsbury, director of chemical research for American Cyanamid, said that one simply could not "conceive of an industrial research laboratory of any magnitude that does not use" radioisotopes. Each lab of this kind required access to a disposal facility. Dr. Leonard M. Freemon, the chief of nuclear medicine at New York's Montefiore Hospital predicted that there would be "a shutdown of nuclear medical facilities throughout the country" (*Business Week* 1979).

The Successful Push for Devolution of Disposal Responsibility

The initial response of the federal government was to federalize LLRW disposal. The IRG initially proposed that the federal government develop and implement a national strategy for LLRW disposal (Greenwood 1982). The NRC and DOE suggested that while new national sites were being identified, LLRW should be shipped to temporary disposal sites on the grounds of existing national laboratories (O'Toole 1979). FEMA pushed for an agreement with state governors that would allow the establishment of interim facilities in several states (*Business Week* 1979).

However, because of a history of federal regulatory failures, the states vehemently rejected the idea of federal control over LLRW disposal. This opposition was best explained by Representative Butler Derrick of South Carolina: "It is my firm belief that the Federal government should not acquire land or run low-level waste sites. The track record of the Federal government in the area of nuclear waste does not lead me to believe we can solve the problem better or faster than the states" (U.S. House Subcommittee 1979b, 5–6). According to a 1989 U.S. Congress Office of Technology Assessment report, "States were convinced that they were better qualified than the Federal Government to assure the protection of their citizens and the environment. . . . [S]ubsequent revelations have confirmed that many Federal facilities have not taken adequate care of nuclear and hazardous materials in the past. . . . States wanted to be involved in decisions regarding siting, technology selection, operator choice, regulation, fee schedules, and public participation." Governor Riley of South Carolina argued that the states were seeking a partnership with the federal government that rejects "the arbitrary imposition of federal will" on decisions regarding nuclear waste (1982, 10).

Riley led a special task force of the National Governors Association (NGA) on LLRW, which referred to the federal approach to LLRW disposal as the "shove-it-down-their-throats technique" (Ivins 1980, 16). The NGA recommended that states assume responsibility for commercially generated LLRW (NGA 1980). President Carter rejected the IRG proposal to federalize LLRW disposal and appointed Riley to chair a newly created State Planning Council (SPC) on nuclear waste issues. The SPC also recommended state responsibility for LLRW. The National Conference of State Legislatures issued a declaration that "primary responsibility for the management of low-level radioactive waste rests with the states" (NCSL 1980, 30). Ultimately, President Carter's radioactive waste management plan, unveiled in February 1980, declared that "states have not played an adequate part in the waste management planning process" (1980, 220). He endorsed the NGA's plan for devolving LLRW responsibility to the states. All nuclear waste policy bills generated in 1980 advocated this devolution of responsibility.

PASSAGE OF THE LOW-LEVEL RADIOACTIVE WASTE POLICY ACT OF 1980

Congress initially considered LLRW disposal policy as a small part of various comprehensive nuclear waste management bills. High-level radioactive waste commanded nearly all of the attention in congressional hearings and debates. The central issue of contention regarding all nuclear waste policy in 1980 was the role of states in determining whether nuclear waste would be stored within their borders. Environmentalists and state government officials wanted a strong state veto provision over the siting of a high-level radioactive waste repository. Representative Thomas Petri (R-WI) bluntly summarized this position: "The public and local officials simply will not trust the Federal Government, and the Department of Energy in particular, not to try to ram down their throats something potentially injurious to them" (*Congressional Record* 1980c, 31934). Petri argued that the states would not support any facility siting if they were not offered the opportunity to reject the ultimate siting decision.

Drew Diel of the Sierra Club testified that it would be impossible to successfully implement new radioactive waste sites if states were denied a significant decisionmaking role. James Cubie of the Union of Concerned Scientists argued that "the Federal juggernaut can get rolling in these processes and if the state has the ability to demand all the information and all the facts, because it has a right to say no, you are going to have a better decision made" (U.S. House Subcommittee 1980, 20, 44). Several congressmen brought up past abuses of federal authority during the Lyons, Kansas, siting attempt and the mismanagement at West Valley, New York.

The nuclear industry opposed the idea of a state veto, however. Sol Burstein of the American Nuclear Energy Council argued that state approval "simply will not work" because no state will want to host a high-level radioactive waste facility (U.S. House Subcommittee 1980, 49). In the end, all compromise efforts concerning the role of the states in the siting process failed, and the future of the legislation was put off until the 97th Congress.

In December 1980, when it was clear that comprehensive nuclear waste policy legislation had failed, South Carolina governor Riley announced that he would close the Barnwell facility to all out-of-state waste if Congress did not pass a bill that at least dealt with LLRW. Representative Butler Derrick of South Carolina championed the cause, lobbying members of the House and Senate for more than nine hours on December 13. Congress passed the Low-Level Radioactive Waste Policy Act in the waning hours of this lame-duck session.

The LLRW legislation never drew much attention. Two hearings were held, with just over a dozen witnesses. The hearings confirmed support for state responsibility over LLRW. As Derrick said on the House floor, "I know of no disagreement or opposition to the concept of this measure" (*Congressional Record* 1980a, 34131). Although there was virtually no input on the issue from environmental groups, the Sierra Club and the Environmental Policy Center supported the LLRW provisions because they felt that state authority would enable greater public participation in the siting process (Resnikoff 1982). The nuclear industry also supported state responsibility, perhaps because, relative to high-level waste, they did not think that LLRW facilities would be difficult to site. In the wake of a year of controversy over the role of states in high-level radioactive waste disposal, the LLRW issue seemed like a modest measure on which all interests could agree.

The hasty debate on the floor of the House and Senate reveals that some members did not fully understand the heterogeneous nature of the LLRW waste stream. For example, when a senator asked for clarification on the nature of LLRW, a bill sponsor explained, "What we are dealing with here is non-defense, non-nuclear power, non-R & D activities. It only deals with the commercially generated waste ordinarily of the kind associated with radioactive development and radioactive waste in hospitals, the radioactive elements that might be involved in independent experiments in laboratories at universities" (*Congressional Record* 1980b, 33965). This statement clearly misrepresents the LLRW waste stream, the vast majority of which comes from nuclear power plants. Senators were also told that this was a "voluntary program," which ignored the fact that states that opted out of regional compacts would eventually be denied access to all facilities. The bill was described in similar terms in the House. Perhaps for this reason, the LLRWPA was passed without any consideration for local opposition, which ultimately was to hamper implementation. There was no mention in either congressional hearings on LLRW or congressional debate on the LLRWPA of possible local opposition to the location of new LLRW disposal facilities.

The LLRWPA offered the first statutory definition of LLRW. However, the definition was not innovative, and LLRW remained a residual category: "radioactive waste not classified as high-level radioactive waste, transuranic waste, spent nuclear fuel, or byproduct material" (42 U.S.C. 2021b). Thus LLRW remained heterogeneous in radioactivity and risk to human health. The LLRWPA established that each state was responsible for providing disposal capacity for the LLRW commercially generated within its borders. Each state could either develop its own LLRW disposal facility or enter into a regional compact to share disposal responsibility. The act included a provision that would enable regional compacts to exclude LLRW shipments from unaffiliated states after 1986. This exclusionary provision was designed as an incentive for states to quickly join regional LLRW compacts.

On December 13, 1980, Representative Morris K. Udall (D-AZ) took the floor of the House of Representatives to support passage of the Low-Level Radioactive Waste Policy Act. "We put the responsibility squarely on the States," he said, "and they want that responsibility" (*Congressional Record* 1980b, 34131). Udall's statement captured both unusual aspects of this legislation. First, the federal government was willing to yield a portion of its authority over nuclear policy to the states. Second, state governments wanted to accept responsibility for low-level radioactive waste disposal.

CONCLUSIONS

By 1980, the states had long since ceased to trust the federal agencies charged with regulating nuclear technology. For decades, the AEC and its successor agencies had dodged the regulatory responsibility they were ordered to execute. This administrative behavior stemmed from the contradictory nature of the Atomic Energy Act and subsequent amendments. The act entrusted a single agency, the AEC, to both promote and regulate nuclear energy production. The AEC aggressively pursued the promotion of commercial nuclear energy at the expense of its regulatory responsibilities. Regulatory neglect led to nuclear accidents, mismanagement, and internal conflict among technical experts. By the 1970s, state governments had become frustrated with federal blunders over regulatory areas typically run by states, such as public health, safety, transportation, waste disposal, and environmental quality.

The Three Mile Island accident in 1979 was simply more evidence of federal mismanagement of nuclear policy. In the wake of the accident, and in the context of expanding environmental regulation and calls for devolution, the states demanded more responsibility over the location and management of all types of radioactive waste. The federal government was not willing to cede significant authority over the disposal of spent nuclear fuel and reprocessing wastes to the states. However, the states did persuade the federal government to yield authority over the disposal of LLRW.

Passage of the resulting LLRWPA was remarkably uncontroversial, reflecting a shared perception of the problem by federal and state government actors. This policy "solution" was designed to address a supply problem of limited waste disposal capacity, a technical problem of safe disposal facilities, an authority problem regarding decisionmaking over new facilities, and an equity problem among states hosting facilities. The LLRWPA sought to dramatically expand waste disposal capacity by shifting authority for the technical issue of establishing a multitude of new facilities to the states. The greater number of sites across the states would improve equity among states in shouldering the waste disposal burden. This solution did not address the generation of LLRW. It did not differentiate among LLRW physical, chemical, or risk characteristics, nor among LLRW producers. The LLRWPA simply sought to establish a vast network of disposal sinks distributed broadly across the United States to facilitate continued generation.

The implications of this approach emerged over decades of failed implementation efforts. As the states sought to establish new LLRW facilities, they encountered alternative definitions of the LLRW problem asserted by the local communities named as candidate sites. Congress had given no consideration in the LLRWPA hearings or debates to the possibility of local opposition to LLRW disposal sites. Likewise, witnesses representing state interests and environmental groups failed to address any local concerns. Even when Congress more thoroughly considered LLRW disposal in the fall of

1985 and amended the LLRWPA, no significant discussion of local opposition emerged. Yet the actual LLRW siting processes took place within the boundaries of municipalities, where both public and local governmental opposition was significant and effective.

CHAPTER 3

GLOWING RECOMMENDATIONS: NIMBY, ENVIRONMENTAL JUSTICE, AND THE FRAMING OF LLRW SITE SELECTION

The LLRWPA solution championed by the states and codified by the federal government reflected not only a shared definition of the LLRW problem based on the perceived need for greater disposal capacity equitably distributed across the states, but also a particular understanding of state government power. The National Governors Association (NGA) and other state actors lobbying for the LLRWPA had convinced Congress that state governments had the appropriate resources to successfully implement what was perceived as a highly technical task. The LLRWPA assumes that states have the power to establish new LLRW facilities. Ultimately, any new facilities would be built within state boundaries, and states possessed the financial resources, organizational infrastructure, and technical expertise to select and build them, as well as the legal authority to impose them on local communities. U.S. constitutional law dating back to the mid-19th century, under a doctrine known as "Dillon's rule," holds that states can create, destroy, abridge, and control local units of government. By law, the states certainly seemed to hold Dahl's classic "intuitive" notion of power over local communities—the ability to get someone to do something he or she would not otherwise do (1969, 2).

The simple conception of state government power reflected in the LLRWPA left this law ill equipped to deal with more complicated dimensions of power relationships exercised among states, between states and the federal government, and between states and local communities. Despite the states' enthusiasm as a collective to wrest LLRW disposal authority away from the federal government, the prospect of a new LLRW disposal facility in any particular state was still perceived as a concentrated cost to be avoided as long as possible. Hayden and Bolduc (1997) pointed out that unlike laws establishing interstate compacts for things such as river management, taxation, or education, the LLRWPA did not establish an equitable sharing of costs and benefits among member states. States that joined regional compacts jockeyed with each other to avoid being the first to host a new site.

At the same time, states that were left out of the compacts fouled up all progress by working to block congressional approval of compacts altogether. Hill and Weissert (1995) argued that implementation of the LLRWPA fell prey to the "irony of delegation." These authors rightly noted that the LLRWPA was not an abdication of authority, but rather the creation of a principal-agent relationship, with Congress, as principal, maintaining oversight of implementation through the approval of compacts. As principal, Congress also maintained the authority to alter the directives given to the states by amending the policy. Hill and Weissert explained that states had an incentive to stall implementation of the LLRWPA because the costs to them of failing to commence LLRW site selection were far lower than the expected value of potential alterations Congress could make to the LLRWPA by failing to approve compacts or otherwise amending the law. The irony of delegation in this case was that it encouraged individual states to ignore the initial implementation directives of the LLRWPA, while waiting for policy alteration from Congress.

As the 1986 target for newly constructed sites approached, the states had failed to act. In 1985, Senator James A. McClure from Idaho said in his opening remarks to the Senate Subcommittee on Energy Research and Development that "no new regional low-level waste disposal facilities are even close to becoming a reality ... states have been exceedingly sluggish in coming to terms with their responsibilities" (U.S. Senate Subcommittee 1985, 44). Few states had even initiated a site selection process.

The states, led by those hosting existing LLRW facilities, asked the federal government to impose a harsher incentive structure. In 1985, the NGA hosted a series of meetings for state government representatives to offer solutions to this collective-action problem. The states persuaded Congress to extend access for all states to existing LLRW disposal sites until 1992, while establishing strict milestones for new LLRW site construction, enacting harsh penalties for noncompliance, and escalating surcharges for disposal at existing facilities in the years prior to 1993. The sharpest teeth in what would become the Low-Level Radioactive Waste Amendments Act (LLRWPAA) required states that failed to provide disposal by 1993 to take legal title to and assume liability for all commercial LLRW generated within their borders.

The strengthened incentives in the LLRWPAA spurred states and regional compacts to begin site selection processes in earnest. Between 1986 and 1993, when a Supreme Court decision would change the policy context once again, states and regional compacts identified more than 20 candidate counties for LLRW disposal. However, just as it seemed Congress had helped remedy intergovernmental difficulties among states, a new intergovernmental difficulty took hold—local opposition to state siting efforts. The following three chapters explore the dynamics of local responses to proposed LLRW facilities in 21 counties.

CASE SELECTION: CANDIDATE SITES FOR AN LLRW DISPOSAL FACILITY

States and interstate compacts began selecting actual candidate sites for LLRW disposal facilities in U.S. counties in 1986, after the LLRWPAA provided more rigorous incentives and firm deadlines. The U.S. Supreme Court decision of *New York v. United States* (505 U.S. 144, 112 S.Ct. 2408, 120 L.Ed. 2d 120 (1992)) nullified the most potent incentives of the LLRWPAA in 1992. Between 1986 and 1992, 25 U.S. counties found themselves on a short list of candidate sites for an LLRW disposal facility. (Candidate site selection in Maine was in process before the *New York* decision, and the sites were named shortly after the decision.) During the late 1980s, 17 of these counties were located in states that had joined an interstate LLRW compact; the remaining 8 were in unaffiliated states (Table 3.1).

Two compacts and 14 states are excluded from this list. The Northwest Compact (with member states Alaska, Hawaii, Idaho, Montana, Oregon, Utah, and Washington) chose the already existing Hanford LLRW disposal site in eastern Washington. This site was established in 1965, during a different era of nuclear politics than the cases listed in Table 3.1. The Appalachian States Compact (with member states Delaware, Maryland, Pennsylvania, and West Virginia) selected Pennsylvania as the host state but did not select actual

Compact name and state membership	Host state	Candidate county
Southeast: North Carolina, Alabama, Florida, Georgia, Mississippi, Tennessee, South Carolina, Virginia	North Carolina	Richmond Rowan Union Wake
Central: Nebraska, Oklahoma, Louisiana, Kansas, Arkansas	Nebraska	Boyd Nemaha Nuckolls
Rocky Mountain: Colorado, Nevada, New Mexico, Wyoming	Colorado	Montrose
Central Midwest: Illinois, Kentucky	Illinois	Clark Wayne
Midwest: Michigan, Indiana, Iowa, Minnesota, Missouri, Ohio, Wisconsin	Michigan	Lenawee Ontonagon St. Clair
Northeast: Connecticut, New Jersey	Connecticut	Hartford Tolland
Southwest: California, Arizona, North Dakota, South Dakota	California	San Bernardino Inyo
None	Maine	Lincoln Penobscot Somerset Waldo
None	New York	Allegany Cortland
None	Texas	Hudspeth
None	Vermont	Windham

Table 3-1. Counties Selected as Candidate Sites for LLRW Disposal Facilities, Their Host States, and Compact Affiliation

Source: This information was drawn from the annual U.S. Department of Energy reports to Congress on LLRW management progress, from 1986 to 1995

candidate sites. Finally, the unaffiliated states of Massachusetts, New Hampshire, and Rhode Island failed to name candidate sites for LLRW disposal.

The LLRW Policy Amendments Act of 1985 required the Department of Energy (DOE) to track state progress toward facility siting. The DOE established the following scale of milestones en route to siting an LLRW facility: (1) create a siting procedure; (2) select candidate sites; (3) characterize candidate sites; (4) select site; (5) complete environmental assessment; (6) submit license request; (7) achieve license approval; and (8) provide disposal. Each of the counties listed in Table 3.1 qualifies as a DOE-recognized candidate site under milestone 2. Each county hosted at least one site that state siting agencies and contractors initially characterized as physically appropriate to host an LLRW disposal facility. The state agencies charged with choosing an LLRW site considered geology, hydrology, population density, and sometimes other factors, such as state ownership of the land in question. The interpretation of these general factors varied from state to state, as did the methods of characterization. Each of these counties passed an initial screening process and then ended up on a short list of candidate sites.

I have included 21 of these 25 counties in this study but excluded the others for the following reasons: in Maine's Somerset and Penobscot Counties, numerous candidate sites had no recorded human population in the vicinity; in Lincoln County, Maine, the local paper did not cover the issue at all; and in Inyo County, California, the site was dropped immediately because of technical shortcomings.

Making a Short List: How Siting Authorities Chose Candidate Counties

How did these counties come to be candidates for LLRW facilities? Although there was no consideration of the intergovernmental dynamics between local and state actors during the passage of either the LLRWPA or subsequent amendments, the selection of candidate counties that began in 1986 revealed that state siting authorities recognized and even feared the potential power of local opposition. By this time, local opposition to everything from halfway houses to hazardous waste was widely recognized as a potent political force. Even before passage of the LLRWPA, the U.S. Environmental Protection

Agency (EPA) had recognized the challenge of local public opposition to controversial facilities. In a national report on hazardous waste facility siting, EPA recommended a "low-profile" approach to reduce the "potential for opposition," noting that "when the public is unaware of a siting attempt, they are unlikely to oppose it" (1979, 18).

Hazardous waste became one of the most prominent news stories of the decade with the revelation that a small working-class neighborhood in Niagara Falls, New York, was sitting on top of a hazardous waste site called Love Canal. There was at least one national television news story a week on Love Canal or hazardous waste for two full years, often with images of citizens struggling against government and industry (Szasz 1994). In 1981, by identifying "locally undesirable land uses," such as hazardous waste sites, as LULUs, Frank Popper (1981) was just the first of many observers to pen clever and increasingly pejorative acronyms for local opposition. These included NOOS (not on our street), NOPE (not on Planet Earth), CAVE (citizens against virtually everything), and the still-popular NIMBY (not in my backyard) (Dear 1992). This last acronym became the most pervasive. Academics studying waste policy picked up on the term and enriched it with telling metaphors, presenting NIMBY as a syndrome to cure (Portney 1991), a new feudalism to overcome (Dear 1992), a dragon to slay (Inhaber 1998), or an obstruction to remove (O'Hare and Sanderson 1993).

In 1988, *New York Times* reporter William Glaberson traced the usage of NIMBY to business executives employing a "mocking nickname" for local opponents of new developments, whom they believed "could push the country toward an unprecedented economic paralysis." He reported that industry had been busy in the 1980s devising ways to defeat the "NIMBY commandos." Glaberson published the results of a 1984 report prepared by Cerrell Associates for the California Waste Management Board, predicting the demographic groups least and most likely to be resistant to a waste facility. The report revealed that for state agencies and private industry charged with siting developments like LLRW facilities, local public opposition had become part of the problem definition, and schemes to avoid this opposition were programmed into the implementation of site selection. Cerrell advised its state client that "constructing a demographic profile" would "assist in selecting a site that offers the least potential of generating public opposition" (1984, 29–30).

Among other characteristics, the least resistant community profile, according to Cerrell, was a small rural population with stable residency, a relatively low median household income, a majority of residents having no more than a high school diploma, and a voting record that was Republican and conservative.

Interestingly, the candidate counties for LLRW facilities chosen between 1986 and 1992 seem to closely match the Cerrell profile of least resistant communities. The average proportion of the population that lived in rural areas in these 20 counties was 65.9 percent, more than 40 percent greater than the national percentage. At $25,979, the 1990 median household income across these counties was over $4,000 less than that of the country as a whole. Across these counties, the education level of an average of 61.8 percent of the population 24 years or older was a high school diploma or less—a figure 7 percent greater than the national average. The candidate counties tended to support Republicans in U.S. House of Representatives elections, averaging a 56.3 percent Republican vote in elections preceding the siting processes.

When the demographics of each county are compared with those of its home state, the same pattern seems to follow (see Table 3.2). U.S. Census data from 1990 show that more than two-thirds of the candidate counties had smaller populations than the county average in their states; 18 of the 21 counties were more than 50 percent rural; 18 of the counties had a higher percentage of people with no more than a high school diploma than their home states as a whole; all but 2 counties had more stable residencies than their states, with a greater percentage of people occupying the same houses for the past five years; and all but 2 counties had lower median household incomes than their states. Political ideologies across these regions are difficult to draw during this era, but voting records reveal that two-thirds of the counties cast a higher percentage of ballots for conservative Republican presidential candidate Ronald Reagan in 1984 than was the average across their home states.

The idea of demographic profiling puts a sociopolitical twist on what had been understood as a technical approach. Scholars have characterized this approach as technocratic (Dahl 1989; McAvoy 1999) or managerial (Williams and Matheny 1995), whereby technical experts implement policy according to criteria that are perceived to be objective. For example, New York's siting authority was required to identify "technically suitable" sites and "demonstrate

County	Population > county average in home state	Rural population > 50%	% high school education or less > state	% of pop. in same house as in 1985 > state average	Median household income < state	% vote for Reagan in 1984 > state
San Bernardino, CA	No	No	Yes	No	Yes	Yes
Montrose, CO	Yes	Yes	Yes	Yes	Yes	Yes
Hartford, CT	No	No	Yes	Yes	Yes	No
Tolland, CT	Yes	Yes	No	Yes	Yes	Yes
Clark, IL	Yes	Yes	Yes	Yes	Yes	Yes
Wayne, IL	Yes	Yes	Yes	Yes	Yes	Yes
Waldo, ME	Yes	Yes	Yes	Yes	Yes	Yes
Lenawee, MI	Yes	Yes	Yes	Yes	Yes	Yes
Ontonagon, MI	Yes	Yes	Yes	Yes	Yes	No
St. Clair, MI	No	Yes	Yes	Yes	Yes	Yes
Boyd, NE	Yes	Yes	Yes	Yes	Yes	Yes
Nemaha, NE	Yes	Yes	Yes	Yes	Yes	Yes
Nuckolls, NE	Yes	Yes	Yes	Yes	Yes	No
Allegany, NY	Yes	Yes	Yes	Yes	Yes	Yes
Cortland, NY	Yes	Yes	Yes	Yes	Yes	Yes
Richmond, NC	Yes	Yes	Yes	Yes	Yes	No
Rowan, NC	No	Yes	Yes	Yes	Yes	Yes
Union, NC	Yes	Yes	Yes	Yes	No	Yes
Wake, NC	No	No	No	No	No	No
Hudspeth, TX	Yes	Yes	Yes	Yes	Yes	No
Windham, VT	No	Yes	No	Yes	Yes	No

Table 3-2. Demographic and Political Characteristics of Candidate Counties for LLRW Facilities

that no obviously superior alternatives can be identified" (National Research Council 1996, 35–36). To identify technically suitable sites, the authority was set up to map, model, and score each region of the state according to "natural site features" such as geology, groundwater hydrology, meteorology, climatology, and ecology (NYSLLRWSC 1989, 1). Siting agencies in other states emphasized the technical nature of site selection. In North Carolina, a siting representative explained that site selection was based on "an examination of technical information by technical professionals using technical criteria" (Rosenberg 1991).

However, fear of local opposition inspired a different sort of process or a different understanding of technical expertise. Siting authorities turned to people like Robert E. Leak to conduct demographic profiles. In an interview in 2003, Leak told me he was an "industry hunter," working "to recruit new industries and businesses" to communities across his home state of North Carolina. Leak assessed the likelihood of opposition to an LLRW facility in counties based on "general socioeconomic and demographic information" and "political profiles" (Epley 1989, I000997). Leak said that he and others took part in "windshield" surveys of the state, where they traveled by car looking for places that were unlikely to oppose construction of an LLRW disposal facility. Notes from these surveys were later released during court proceedings challenging the site selection. One consultant wrote comments about certain sites visited during the windshield survey such as "trailers everywhere," "distressed county," or "very depressed area," then found undesirable physical characteristics such as sandy soil or marshy wetlands, and still summed up the site as "in" rather than "out." Conversely, notes next to other counties that were awarded "out" status read "affluent" or "economic development" (Farren 1992).

Information like this was later summarized in a confidential public relations assessment by Epley Associates. The report focused exclusively on socioeconomic and political factors thought to distinguish certain potential sites as less likely to form active opposition. In general, the authors of the report tended to favor sites that were suffering from economic hardship. For example, they characterized one county that would become a candidate site in the following way: "Relatively poor and undeveloped, Richmond County would benefit immensely from the economic rewards of hosting the low-level radioactive waste facility. ... Given Richmond County's less-than-thriving economy and the lack of environmental activism within the county, siting the facility might be among the less difficult of the potential sites." The authors also noted that "the county has no apparent history of environmental controversies or environmental activism" (Epley 1989, 6352). North Carolina state representative George Miller, who sponsored the original LLRW legislation and was intimately involved in the siting process, told me in a 2003 interview, after his retirement, that "the sites were not primarily selected or removed on scientific grounds; the site selection was political—they wanted to find sites that would not object."

In the New York siting process, a National Research Council review found that "[technical] performance and socioeconomic criteria were combined inappropriately during Candidate Area Identification and Potential Site Identification screening." The reviewers argued that this faulty practice enabled some sites with poor technical characteristics but favorable socioeconomic characteristics to garner a high enough total score to advance to candidate site selection. The reviewers also found that a site that was volunteered by a landowner was named as a candidate site even though it did not meet the minimum cutoff score applied statewide. This violated the state regulation that required candidate sites to be at least as good as all other sites in the state. The reviewers found that the volunteered site in question should have been disqualified because of unsuitable soil composition (National Research Council 1996, 130, 145). The reviewers' conclusions match interview statements obtained from the landowner volunteering the site. In a 2002 interview, the landowner, Art Allen, described the commission's reaction to his proposal to sell his farm in the following way: "They'd already picked their five sites, and one of them just kind of glanced it over and they said, 'This is just what we've been looking for,' so they bumped the fifth site and put us on the list, and that's how we come to get on it. It was just a spur-of-the-moment thing." This occurrence typifies a flawed process that improperly inserted socioeconomic factors and private property ownership into the search for technically suitable sites. The National Research Council reviewers argued that the New York State LLRW Siting Commission "did not always follow its own procedures as defined in the Siting Plan" (National Research Council 1996, 130).

One final example from Illinois reveals that the state's Department of Nuclear Safety (DNS) determined that at least one candidate site was suitable before the experts it hired to conduct this technical analysis had completed their reports. An Illinois senate investigation found that Terry Lash, the director of the DNS, told personnel involved in the siting process that "site screening is more of an economic consideration than a safety consideration" (Illinois Special Counsel 1990, 8). Lash declared that safety would not be addressed by the DNS until after economic analysis of possible sites.

The Martinsville site in Clark County, Illinois, quickly became a favored site for nontechnical reasons, including economic factors, local political support, and access to a major interstate highway. However, the Martinsville area also had less-than-desirable groundwater hydrology characteristics. Every independent scientist working on the project, including contractors from

Westinghouse, Earth Technology Corporation, and Battelle Corporation, agreed that the site sat on top of an aquifer and needed additional study. In spite of this evidence, Lash testified before the Illinois senate that the site was, in fact, not located above an aquifer and that "the facility could not conceivably pose a threat to the drinking water supplies of Martinsville." Lash then directed independent contractors to prepare a statement "that explains why the Martinsville Alternative Site is considered to be a 'technically excellent site.'" In the opinion of the contractors, Lash "sought to compromise the professional integrity of its independent consultants by directing which scientific conclusions they should reach" (Illinois Special Counsel 1990, 13, 15). The contractors refused to issue the report. The Senate Special Counsel ultimately chastised Lash and the DNS for their "premature and ambiguous insistence that Martinsville was 'technically excellent.'"

This evidence from these three states that underwent independent reviews or court proceedings concerning the LLRW siting process indicates that demographic characteristics played a role in candidate site selection in ways often at odds with the technocratic language of the site selection guidelines and the siting authorities themselves. Of course, as Fischer has noted, "In the 'real world' of public policy there is no such thing as a purely technical decision" (2000, 43). Even when "perfectly" applied, a technocratic approach must engage uncertainties and trade-offs among alternatives that necessitate value-laden decisions—despite the use of technical criteria employed by trained experts (Dahl 1989). As a result, a technocratic approach is exceedingly vulnerable in application. Any deviation from the technical criteria, or any indication of a value judgment on the part of the "experts," could undermine public trust in the process and the authorities charged to implement it. Demographic profiling in site selection had just this effect. Such profiling was an attempted implementation "solution" that followed from the inclusion of local opposition as part of the LLRW problem definition. But the perception of injustice in these practices became the cornerstone of an alternative problem definition put forth by local opponents—the problem of environmental justice.

In the early 1980s, just as industry groups were popularizing the NIMBY acronym to describe local opposition, the activists comprising that opposition were popularizing the concept of environmental justice. In 1982, local activists joined with national leaders of the civil rights movement in acts of civil disobedience to oppose the construction of a hazardous waste facility in Warren County, North Carolina. The activists argued that the site was technically inferior because of its shallow water table and that state authorities had targeted Warren County because it was home to a higher percentage of low-income and black residents than the rest of the state. Despite marches, hunger strikes, and sit-ins in front of dump trucks, the site was successfully completed. However, the link between the siting of waste facilities and justice concerns gained momentum. Highly publicized studies by the U.S. General Accounting Office (GAO 1983) and the United Church of Christ's Commission on Racial Justice (UCC 1987) found that hazardous waste sites were disproportionately located near minority communities. In 1991, the National People of Color Environmental Leadership Summit convened a multitude of community activists in what could be fairly called a social movement around shared policy goals. A new field of academic analysis focused on the demographic distribution of existing sources of negative environmental consequences (Bullard 1983; Clarke and Gerlak 1998; Mohai and Bryant 1992; Mohai and Saha 2007; Timney 1998; Allen et al. 2001; Goldman and Fitton 1994) and the development of an environmental justice movement (Bullard 1994; Foreman 1998; McGurty 1995; Novotny 1995; Pellow and Brulle 2005; Pulido 1996; Roschke 1997; Steinberg 1999; Szasz 1994). In 1994, President Clinton signed Executive Order 12898 "to address environmental justice in minority populations and low-income populations." EPA developed the following definition of environmental justice: "The fair treatment and meaningful involvement of all people regardless of race, color, national origin, or income with respect to the development, implementation, and enforcement of environmental laws, regulations, and policies. Fair treatment means that no group of people, including racial, ethnic, or socioeconomic group should bear a disproportionate share of the negative environmental consequences resulting from industrial, municipal, and commercial operations or the execution of federal, state, local, and tribal programs and policies" (EPA 1998, 6).

The EPA guidance document on environmental justice defines two populations of concern: "minority populations," which are taken to mean all self-identified nonwhite people and Hispanics; and "low-income populations," which include those households that fall below the annual statistical poverty thresholds established in the Current Population Reports of the U.S. Census Bureau. EPA's guidance is less clear on the thresholds of minority and low-income populations that should raise environmental justice concerns. It suggests that a population may be significant if it is "meaningfully greater" than the minority or low-income population in the general population or other "appropriate unit of geographic analysis" (EPA 1998, 11–12). Radion International, a private contractor that consults on environmental justice compliance issues, developed a formula of relative ratios, comparing the demographics of the immediate area hosting a given site to the larger area out of which the immediate site was selected (Crum et al. 1999). Following is the formula for the relative ratio strategy for environmental justice analysis:

$$R = \text{relative ratio} = p1/p2$$
$$p1 = (T_p - NM_p)/M_r$$
$$p2 = (T_p - M_p)/NM_r$$

where p is the proximate zone in the area within a given region of influence (in this case, the county), r is the reference area outside of the region of influence (in this case, the state minus the county), T is the total population, M is the minority population, and NM is the non-minority population. The same formula holds for low-income assessment, with the substitution of low-income population for minority population and non-low-income population for non-minority population.

This strategy applies the rule that if the relative ratio is less than 1, there is no potential for disparate environmental harm. Table 3.3 presents the relative ratios for minority and low-income populations in the 21 counties facing a proposed LLRW facility.

The results of this assessment reveal that most counties selected as candidate sites for an LLRW facility were predominantly poor and white relative to their home states. There were just four counties with a relative ratio of 1 or greater in the minority column: Hartford, Connecticut; Richmond, North Carolina; Hudspeth, Texas; and Windham, Vermont. Hudspeth County has a very large score in this column, generated by its high minority population relative to the state of Texas. The most remarkable result is that 14 of the 21 counties score a relative ratio greater than 1 in the low-income column.

County	Minority relative ratio	Low-income relative ratio
San Bernardino, CA	0.86	1.02
Montrose, CO	0.58	1.43
Hartford, CT	1.41	0.64
Tolland, CT	0.31	0.41
Clark, IL	0.02	1.59
Wayne, IL	0.03	1.87
Waldo, ME	0.53	1.37
Lenawee, MI	0.42	0.81
Ontonagon, Mi	0.08	1.81
St. Clair, MI	0.22	0.95
Boyd, NE	0.13	2.04
Nemaha, NE	0.18	1.37
Nuckolls, NE	0.06	1.48
Allegany, NY	0.05	1.18
Cortland, NY	0.05	1.04
Richmond, NC	1.35	1.37
Rowan, NC	0.62	0.99
Union, NC	0.63	0.69
Wake, NC	0.95	0.59
Hudspeth, TX	3.12	1.74
Windham, VT	1.03	1.13

Table 3-3. Relative Ratio Scores for Counties Facing a Proposed LLRW Facility

NIMBY OR ENVIRONMENTAL JUSTICE?
FRAMING THE SITE SELECTION PROCESS

The exercises of determining whether the information above on demographics of candidate counties for LLRW facilities and the use of such "nontechnical" characteristics in site selection processes are evidence of environmental injustice, and of assessing whether local opposition that emerged in these counties constitute a NIMBY response, are not as essential to understanding implementation of the LLRWPA as is a grasp of what the dominant public perception of these siting processes came to be. NIMBY and environmental justice are not objective descriptors, but subjective and politically strategic frames of meaning reflecting competing definitions of the LLRW policy problem. Scholars have employed the term *framing* to help understand how political actors strategically create messages in an attempt to further their cause (Snow et al. 1986).

The NIMBY frame advanced the LLRW problem definition that federal and state decisionmakers came to share, centered around continued waste generation, the technical development of greatly expanded disposal capacity, and equity in siting across states. The LLRWPA solution to this problem assumed that state authority would select and, if necessary, impose upon local communities to host new LLRW facilities. The NIMBY frame advances this problem definition and solution. The NIMBY phenomenon is generally defined as the refusal of local populations to host the negative environmental consequences of a larger public good (Wolsink 1994). After reviewing the academic literature on NIMBY, Kraft and Clary summarized the NIMBY characterization of local opposition as "poorly informed, interested primarily in avoiding local imposition of risks, and emotive rather than cognitive" (1993, 96). Under this frame, local opposition is a problem that is best avoided, or at least managed to prevent unqualified, self-interested, and emotional local actors from obstructing the technical process of creating new LLRW disposal capacity for the common good.

There is evidence that LLRW siting officials attempted to depict local opposition to proposed LLRW sites with NIMBY characteristics such as ignorance and selfishness. US Ecology, an LLRW contractor in Nebraska, wrote the following warning in a newsletter to people in the candidate county of Boyd: "There is no question that there are individuals and groups, who because of their opposition to this project, have a vested interest in prejudging the sites and presenting their opinions as fact. ... The possibility of such conjecture being popularly accepted as fact is prime evidence of why we must make environmental decisions primarily on scientific study rather than perception or politics" (as quoted in Snowden 1997, 117).

This quote depicts Boyd County opposition leaders as acting selfishly out of "vested interest" and ignorantly "prejudging" the site based on opinions. US Ecology argued that Boyd citizens should not be swayed by such "perception" or "politics." Instead, serious environmental decisions such as this should be based on "scientific study." Similarly, Angelo Orazio, chairman of the New York State LLRW Siting Commission, accused activists in Allegany County of treating local residents like "mushrooms." He claimed that activists were keeping residents "in the dark and feeding them manure. ... We have to remember that in spite of all our efforts, many people are suffering from information starvation" (Dickenson 1990).

However, the local political actors themselves had a very different definition of the LLRW problem that challenged the very generation of the waste, the feasibility of disposal technology at new sites, equity across communities, and democratic fairness in the site selection process. The environmental justice frame, with its emphasis on fair treatment, public involvement, and equity at the community level, advanced this problem definition. Once NIMBY and environmental justice are treated as strategic frames, it should come as no surprise that local opposition expressed itself with rhetoric that did not fit the NIMBY characteristics, but instead latched on to environmental justice concerns.

Opponents responded to accusations of NIMBYism by educating themselves on the issue of LLRW disposal. Opponents in Allegany County, New York, responded to Orazio's mushroom depiction with declarations like this: "Get involved. Get Informed ... I do my own research. With access to the pros and cons of nuclear waste, I feel capable of making my own decisions" (Jefferds 1990). Another opponent explained, "What the Siting Commission does not seem to realize is that the citizens throughout these counties and throughout the world have educated themselves to the truth about so called 'low-level' radioactive waste" (Gardner 1990). An opponent in Boyd County, Nebraska, explained the self-education campaign this way: "Well, you know, they [US Ecology] tell us one thing and then they say 'get educated. You'll understand it.' Well, you know the more educated we got, the less we wanted it. They didn't seem to understand that" (as quoted in Snowden 1997, 121).

There is some evidence that LLRW opponents consciously moved away from NIMBY themes and strategically adopted alternative frames. During the course of the siting process, some letters to the editor conveyed messages specifically to active LLRW opponents, urging them to avoid a NIMBY approach to the struggle. For example, in the days leading up to a public hearing on the LLRW siting, the following letter to the editor appeared in the *Richmond County Daily Journal*: "It is important that people from Richmond County show up [to the hearing], show their concern, seek answers to their questions and state reasons why they don't want the facility here. It is also important that those present not get carried away by their emotions. It's an emotional issue, but rational arguments are more helpful for Richmond County's cause. ... Don't make it easy to portray Richmond County opposition to the site as the work of ignorant roughnecks. There are too many good and logical arguments against placing the site here to allow that to happen" (1989d).

This is clearly a message directed at LLRW opponents with the purpose of moving away from the NIMBY characteristics of technically ignorant, personalized, and emotional opposition. During the same week, the *Salisbury Post* in Rowan County, North Carolina, printed a letter to the editor warning that emotional responses were "self-defeating": "If I scream in your face, you tend to remember that I screamed in your face instead of what I said and whether it made sense. ... If you were one of them [the state LLRW siting officials], which would sway you more: emotion and rudeness and mass hysteria? Or an attempt to persuade you, through a rational and respectful marshalling of facts, that another site would be more appropriate?" (Bouser 1989b).

A letter in the *Enquirer-Journal* in Union County, North Carolina, warned that outsiders viewed NIMBY groups "as being interested in self economic and social preservation and not the concerned residents and citizens that they are." The author then suggested that opponents focus on technical issues such as transportation and groundwater (*Monroe (NC) Enquirer-Journal* 1989b). An opponent in St. Clair County, Michigan, chastised fellow opponents for writing letters to the editor that were "absolutely ridiculous, from uninformed people on this issue." The author urged citizens to provide "realistic solutions" rather than "misinformation, rumors and threats" (Bundy 1989). Messages like these demonstrate a strategic effort to adopt a less personal and emotional response with a higher technical level of knowledge.

Opposition

State siting authorities and their contractors did not succeed in avoiding local opposition. Although, as we will see in succeeding chapters, the activities of opponents varied dramatically across candidate counties, opposition was the dominant response to the LLRW issue in letters to the editor of the local papers across these cases. In the analysis that follows, I examine letters to the editor regarding the proposed LLRW facility in the closest daily newspapers to each of 21 counties named as LLRW candidate sites for the duration of the siting process. The duration for each case spans from the day when the county was first publicly announced to be on a short list of candidate sites to the day the

Reaction	Percentage	N
Opposed	83	803
Neutral	6	54
Positive	11	108
Total		965

Table 3-4. Reaction to LLRW Site Proposals in Letters to the Editor
Note: Rounding error is present in the percentages

siting process effectively stopped. Public expressions such as letters to the editor are less an indicator of public opinion than a measure of how the issue was framed for public consumption by the "attentive public" engaged in the LLRW issue.[1] My analysis found that 83 percent of the letters to the editor regarding LLRW expressed opposition (Table 3.4).

Not only was the prevalence of an opposition frame to the LLRW sites overwhelming, but it was strongly expressed. The letters to the editor that expressed opposition to LLRW can be overwhelmingly categorized as "strongly opposed" (Table 3.5).

Finally, letters of support for the LLRW failed to constitute a majority of letters on the issue in any of the 21 candidate counties. Column three of Table 3.6 presents the letters of support as a percentage of total letters on the LLRW issue. The letters to the editor on the LLRW issue were gathered over siting processes with different durations. To control for this difference, column two of the table presents the percentage of days during the siting process in

---

[1] Often scholars addressing the NIMBY issue use survey data gathered from populations in the midst of a controversial siting process (Frey and Oberholzer-Gee 1997; Hunter and Leyden 1995; Kunreuther and Easterling 1990; Summers and Hine 1997) to answer questions such as "How does the public feel?" and "Why do people feel the way they do toward waste facility siting issues?" (Lober 1995, 499). The LLRW disposal issue concerns facility siting processes that have already ended. Although some survey data regarding the LLRW site proposals do exist (Krannich and Albrecht 1995), there is no consistent and comprehensive data source. There also is no comprehensive record of public hearings on the issue of the type Kraft and Clary (1993) coded for high-level radioactive waste disposal. The best and only available information source on public reactions to LLRW across a multitude of counties that faced a site proposal is letters to the editor in the local newspapers. Content analysis of these letters does not measure public opinion, but rather reveals the strategies of those attempting to influence the issue in the media by identifying dominant frames.

Reaction	Percentage	N
Opposed	84	43
Neutral	2	1
Positive	14	7
Total		51

Table 3-7. Reaction to LLRW Site Proposals in Letters to the Editor Authored by Self-Identified Local Government Officials
Note: Rounding error is present in the percentages

The predominant frame on the issue of LLRW as expressed in letters to the editor in candidate counties was one of opposition. If the state siting agencies and contractors were indeed trying to locate willing, or at least unopposed, candidate sites based on socioeconomic characteristics, these results indicate that they failed.

The letters to the editor in candidate counties also reveal that local officials were predominantly opposed to the proposed LLRW sites. Table 3.7 shows that most of the letters penned by elected officials expressed opposition.

In addition, local governments (municipal, county, or both) in 18 of these 21 cases issued a formal statement of opposition to the proposed LLRW site. Even Boyd and Clark Counties, where there was organized support for the facility, issued formal statements of opposition.

Framing the Opposition: Avoiding NIMBY

But how was the opposition framed? It did not easily fit a NIMBY characterization. Kraft and Clary, who summarized NIMBY characteristics in the literature as "poorly informed, interested primarily in avoiding local imposition of risks, and emotive rather than cognitive," developed a three-part categorization scheme to assess the technical awareness, geographic scope of concern, and personal and emotional claims made in public hearings on high-level radioactive waste disposal (1993). I apply this scheme to the letters to the editor on LLRW in the candidate counties. Table 3.8 presents the results of this content analysis.

I found 45 percent (431) of the letters addressed the technical suitability of the LLRW facility. Of these letters, 79 percent exhibited either a high or moderate level of technical awareness, with high awareness characterized by the proper use of scientific terminology or technical criticism of specific aspects of

Technical knowledge	Percentage	N
High	45	196
Moderate	35	149
Low	20	86
Total		431

Table 3-8. Technical Awareness Exhibited in Letters to the Editor on the LLRW Issue
Note: Rounding error is present in the percentages

the LLRW facility, and moderate awareness exhibiting a general understanding of technical aspects of the LLRW facility or criticisms of the suitability of the site without scientific terminology. The remaining 21 percent of the letters addressing technical suitability exhibited a low level of technical awareness, evidencing little understanding of the technical details relevant to suitability.

These letters expressed opposition to the LLRW facility based on the physical characteristics of the site, the properties of the waste stream, and the repository design. Thus they do not easily fit the NIMBY characteristic of an uninformed public. In addition, these findings match those of Kraft and Clary, who found 67 percent of individuals testifying on facility suitability at high-level radioactive waste hearings demonstrating moderate to high levels of technical knowledge (1993, 97).

The geographic orientation of the opposition is measured by determining whether the focus of concern in each letter was exclusively local or expanded to a broader region of concern as well (Table 3.9). The NIMBY response is characterized by an exclusively local concern.

Table 3.9 shows that 87 percent of the letters did convey some local concern; however, only 44 percent exhibited an exclusively local focus of concern. A majority of letters had a scope of concern that extended beyond the local area slated to host the LLRW site. Again, these findings match Kraft and Clary's analysis, which revealed that exclusively local concerns failed to constitute a majority of their responses.[2]

---

[2] It should be noted, however, that this analysis finds a much larger percentage of exclusively local responses—45 percent compared with Kraft and Clary's 23 percent. This could be due to the fact that the high-level waste hearings were statewide in scope.

Geographic focus	Percentage	N
Local	87	836
State	42	406
Other states or nations	11	103
International	2	17
Local only	44	420

Table 3-9. Geographic Focus of Letters to the Editor on the LLRW Issue
Notes: Percentages refer to those demonstrating a given geographic focus. Letters could express multiple foci, and therefore the total exceeds 100 percent.

Emotive themes	Percentage	N
Personalization of LLRW issue	14	134
Emotional threats made against siting authorities	15	144

Table 3-10. Emotive Themes Expressed in Letters to the Editor on the LLRW Issue
Notes: Percentage refers to the percentage of total letters expressing the emotive themes. Letters can fit into each of these categories.

Finally, Table 3.10 shows that personalization of the LLRW issue and emotional claims in the LLRW letters to the editor were fairly rare, constituting just 14 percent and 15 percent of the total letters on the LLRW issue, respectively. These findings also track those of Kraft and Clary.

The point here is not that the local response to the LLRW facilities was either NIMBY or not NIMBY, but rather that—at least, in letters to the editor—it tended to be framed in a way that could not easily be characterized as NIMBY. Taken as a whole, these letters do not reflect technical ignorance, exclusive concern with the local area, or personal and emotional expressiveness.

FRAMING THE OPPOSITION: A SOPHISTICATED INJUSTICE FRAME

If the letters to the editor do not reflect a NIMBY frame, how were the opponents framing their concerns? Opponents predominantly employed a sophisticated injustice frame. As an example of the contrast, the following excerpts from opposition letters in Richmond County, North Carolina, exhibit the hallmarks of a NIMBY response—a low level of technical sophistication, emotional and personal claims, and an exclusive concern with the local area:

> I am opposed to the state putting a low-level radioactive waste dump here in Hamlet. It could sink into the water and kill us all. It could cause all the people to move, or what little we have. Please put the dump somewhere besides Hamlet. (Robertson 1989)

> I don't think that Hamlet is a good place for the site. Hamlet is a small town and people like it. It could kill our families and friends and most of all our dogs or pets that we love so much. Let's fight so that the waste plant site won't come here. (Butler 1989)

> The people of Richmond County do not want a low-level or a high-level nuclear waste dump. ... Our children and grandchildren may be living here someday. What if it leaks into the water supply, it will take hundreds, maybe even thousands of years before it goes away. Nuclear waste is very dangerous and people can die from it. So please think about the consequences that the people of Richmond County have to suffer, because of a nuclear waste dump being placed here. (Griffen 1989)

Letters like this were the exception. Many letters expanded the scope of concern beyond the candidate county. The following letter from Cortland County, New York, illustrates this in the simplest terms:

> I am not in support of the nuclear dump in Cortland County. With that fact in mind, I do not want the dump in other counties either. ... The solution is to stop nuclear power. (Darling 1989)

Many letters attacked the technical credibility of the siting agencies, as in this letter from Boyd County, Nebraska:

> Nebraska Department of Environmental Control title 194—Rules and regulations for the disposal of low-level radioactive waste, chapter 5, Section 001, states: "the disposal site shall be generally well drained and free of areas of flooding or frequent ponding. Waste disposal shall not take place in a 100-year flood plain or wetland as defined in Executive Order 11988, 'Floodplain Management Guidelines.'" The Butte site contains about 40 acres of federally certified wetlands. Ponding typically occurs during periods of normal precipitation. Do you wonder if US Ecology really cares about our safety when they blatantly disregard the rules and

regulations of NDEC? Do you wonder if NDEC is really functioning as a watchdog for the welfare of the people? Do you wonder if the Butte site was picked because of suitability or availability? (Zidko 1989)

Letters like this affix blame to the siting agencies and contractors for failing to do a technically competent job choosing suitable candidate sites and imposing unacceptable risks on the community. Scholars of risk perception have shown that the general public ranks radiation risks from nuclear power and all types of radioactive waste (Slovic 1996). One of the factors these scholars associate with high risk perception in the general public is the degree to which the risk is perceived as involuntary. By attacking the technical credibility of the siting authority on an issue the public is already predisposed to fear as risky, opponents could easily paint the authorities as unjust. These kinds of arguments fit what Snow and Benford have identified as a "master frame" of injustice. Such a frame identifies "some existing social condition or aspect of life and define it as unjust, intolerable, and deserving of corrective action." It also attributes "blame for some problematic condition by identifying culpable agents" (1992).

I found that the majority of letters to the editor (63 percent, or 586 of 931 letters) on the LLRW issue attacked the credibility of siting authorities or contractors. Nearly 85 percent (791 of 931) of the letters charged that the decisionmaking process was unfair, undemocratic, or improperly influenced by political rather technical criteria, often latching on to the very kinds of demographic profiling practices authorities were using to avoid opposition. Opponents depicted their communities as victims of an unjust, technically flawed siting process.

Snowden's analysis of opposition rhetoric in Boyd County, Nebraska, demonstrates how an injustice frame is combined with an attack on technical credibility. She reported that members of the Save Boyd County opposition to LLRW "developed a perception of environmental injustice." She included the following quotation from an activist to exemplify this frame:

They do not care about our safety or whether they kill us. We are a poor rural site, they thought we were ignorant. We are the fifth poorest county in Nebraska, we are expendable. ... This was never sited on technical merit. (1997, 146)

Thomas's 1993 analysis in this same county revealed that the most pervasive theme was that the Boyd County site was a product of politics. Subcategories of this theme included the idea that the siting process was flawed, the siting officials were corrupt, and the need for community consent was promised but not given. He presented the following quotations from opponents as examples:

> [The siting process] appears to be, to me, largely political in the early stages. It appears that a site was chosen with the criteria in mind that it be a long ways away from the press, that it be against a state line ... the people on the other side of the state line, they can't vote. They don't have much access to your media. And, so it really eliminates political problems. (1993, 89)

> They didn't look at anything geologically when they picked that site, I don't believe ... why, (laughs) the water level's two foot below the ground or so. You know, forty some acres of wetlands, and it was just a half-section of land right along the highway that they could buy. I think that's all they looked at. If you look at the county maps, they'll tell you it's not a good building site, even to set a building on, because of the shrink and swell of the ground and the high water table and stuff. (1993, 86)

> The site selection was very poorly done. This site characterization appears to be a conspiracy of the Compact Commission, US Ecology, and the Nebraska Department of Environmental Control to site a dump in an area out of sight, out of mind, with no consideration of guidelines, rules, or regulations. (1993, 84)

> They [the governor's Citizen Advisory Committee] were to study this issue and listen to the concerns of the people. ... This area was never ever represented on this issue. (1993, 111)

These quotes exemplify an evaporation of public trust in the LLRW facility siting processes and a raw sense of injustice that was symptomatic across the candidate counties.

Often the authors of letters to the editor were able to frame opposition in ways that disavowed NIMBY characteristics, questioned the technical competence of the siting process, and leveled a charge of injustice against the siting authorities or government. The following excerpt from a letter in the *Hartford (CT) Courant* hits all of these themes:

> Calling us "NIMBY" is far from accurate. We don't want nuclear waste in anybody's backyard. To be sure, our concerns are about health and safety and

about property values. But listening to residents speak at these meetings reveals the much larger moral issue of freedom itself. The American Revolution was fought because we could not stomach taxation without representation. We certainly did not like the taxes themselves, just as we don't like the ill effects of hazardous waste dumping. But then, as now, we are dealing with a more fundamental principle: that we the people must be considered and must be heard. . . . [T]he selection process claims to be blind. But blindness here does not equate to impartiality; it equates, instead, to not seeing, to a dismissal of values and, ultimately, to a real and figurative darkness. The selection committee did not set foot on the three sites or even in the three towns involved. Instead, the committee relied on evidence obtained from aerial photographs. As Americans we react in the only way we can when our rights are threatened: we protest. . . . States need to join together today to find a common solution to the nuclear waste problem, to find national sites away from homes and schools. But let no one, neither King George III in 1773, nor any superagency in 1991, impose a "solution" without our representation. (Policelli 1991)

Opponents in St. Clair County, Michigan, were able to stitch together a similar message, exemplified in this letter:

The location of a so-called low-level radioactive waste disposal site in our cherished water wonderland may well be the worst of all blunders. Much of the criteria for site selection is already being ignored. But we can't rely on others to support us unless we offer convincing evidence as to why St. Clair County is a bad choice for a disposal site. Fortunately, there are good reasons for all Michigan residents to be concerned about dumping radioactive wastes in St. Clair County. For example, the proposed site's close proximity to Lake Huron and the St. Clair River is a problem. The main link of the Detroit Water Authority's Lake Huron pipeline passes through the site. This pipeline supplies 120 million gallons of water a day to residents of Southeast Michigan, and as far West as Flint. It seems remarkable to me that a team of scientists choosing the candidate areas "didn't know about the Detroit Waterline Oct. 4 when identifying 16,750 acres in Brockway, Emmett, Mussey and Lynn township as a potential home for the proposed site." (London 1989, 6a)

Letters like this show just how vulnerable the site selection processes were to contests over the dominant perceptions of the siting process in the candidate counties. Opposition leaders were easily able to shake the NIMBY label and instead create frames of meaning that discredited both the technical competence and democratic legitimacy of the state siting authorities.

CONCLUSIONS

The LLRWPA reflected a simple assumption of state power to establish new LLRW facilities in local communities, equitably distributed among the states. Once the policy was amended with harsher incentives to force states to begin searching for host communities, intergovernmental dynamics between local- and state-level actors emerged that had not been considered in the design of the LLRWPA.

Although states wielded significant power over local communities, including both the technical expertise and the legal authority to select and impose LLRW facilities on candidate counties, they recognized that local communities could also wield significant power. This was evident generally during the 1980s in the application of NIMBY as a frame to characterize local opposition as part of the policy problem surrounding waste disposal. This was also evident in examples of state LLRW siting authorities employing demographic profiles in hopes of locating communities unlikely to oppose the new facilities.

For more than 30 years, political theorists have written about three dimensions of power (Gaventa 1980). In this intergovernmental conflict between state and local political actors, all three dimensions come into view. The first dimension of power is often described as political resources that can be brought to bear on decisionmaking processes—finances, organization, expertise, votes, authority (Dahl 1969). The LLRWPA assumed that state governments had adequate resources to establish new LLRW facilities. Once implementation began, however, the states seemed to recognize that local communities were not without political resources. Although states had the legal authority to impose the facilities on local communities, the vast majority of elected officials at the federal, state, and local levels depend on the electoral support of local communities. The fear reflected in the concept of NIMBY is the fear of applied local political resources.

The second dimension of power is less visible and includes the "rules of the game" that determine who participates, what is to be decided on, and how the decisions will be made (Bachrach and Baratz 1962). State siting authorities were operating in this dimension of power when they designed and implemented site selection procedures. In fact, in many cases, they may have been trying to minimize the likelihood of local application of political resources in the first dimension of power by using "rules of the game" that selected for demographics thought to be associated with acquiescent communities.

A third dimension of power flows from the strategic construction and communication of meanings by political actors (Gaventa 1980, 15). It is in this dimension that competing problem definitions and strategic frames vie for public adoption. In the LLRW siting process outlined above, local opponents to LLRW facilities worked strategically to avoid fitting the NIMBY frame that siting authorities were applying and instead created a sophisticated injustice frame. This latter frame of opposition seized on the ways siting authorities had strayed from technical criteria in the site selection processes and worked to undermine the legitimacy of the processes by marking the authorities as technically incompetent, politically motivated, and fundamentally undemocratic.

At the very least, the injustice frame of opposition dominated expressions in local letters to the editor. Careful consideration of this frame is important, because social movement scholars recognize this particular aspect of the third dimension of power as a key factor facilitating political mobilization. In the next chapter, I evaluate the relationship between the injustice frame and other factors of active opposition to LLRW site proposals.

CHAPTER 4

POWER GENERATION: ACTIVE OPPOSITION TO LLRW SITE PROPOSALS

What factors account for *sustained active* local opposition to projects such as LLRW facilities? If opposition was the dominant response in letters to the editor across the candidate counties, to what extent did each community act on this opposition? Opposition does not necessarily translate into sustained activism. Both social movement scholars and the siting authorities charged with locating the LLRW facilities recognized the potential power in active local opposition—as Tarrow explains, local communities can muster significant political power when they "join forces in contentious confrontation with elites, authorities and opponents" (1994, 1). However, both scholars such as Tarrow and siting professionals such as Cerrell Associates (1984) have searched for factors to explain variation in mobilization. What McAdam et al. (2001) call the "classic social movement agenda" has sought to identify necessary and sufficient conditions for mobilization, putting forth three broad concepts to account for mobilization: political opportunity structures, mobilizing structures, and collective action frames. As we saw in the previous chapter, consultants working for the hazardous and radioactive waste industry have searched instead for demographic factors associated with levels of local resistance. The 21 candidate counties, each of which faced the same environmental "threat" at roughly the same time, provide a context in which to test these hypotheses. Academic coverage of the LLRWPA has described widespread active opposition to the LLRW site proposals (Albrecht 1999; English 1992). However, the number of collective acts of public opposition varied significantly across the candidate counties. For example, LLRW opponents in Cortland County, New York, averaged one collective act of public opposition every 3 days during the siting process, whereas opponents in San Bernardino County, California, averaged just one act every 96 days.

What accounts for such variation? The quantitative findings testing the hypotheses of both social movement scholars and the siting professionals included in the appendix to this book are not very robust and do not generally support either hypothesis. Nevertheless, they do serve to set up case selection for much more revealing qualitative findings in this chapter. The one factor that does emerge from the quantitative analysis with a significant positive correlation to mobilization is the prevalence of an injustice frame. However, unlike the other variables in the hypotheses, a frame is not so much a static factor as it is a product of dynamic social relationships. The qualitative findings suggest that a search for "factors of mobilization" is misplaced because mobilization *moves*, and the power of local opposition comes from the *movement*. We need instead to understand the dynamics of mobilization, evident in patterns of relationships—social mechanisms that shape just how a community interacts with its political resources.

VARIATION IN COLLECTIVE ACTS OF PUBLIC OPPOSITION TO LLRW SITE PROPOSALS

These candidate counties varied widely in the number of collective acts of public opposition generated against the site proposal. Collective acts of public opposition are defined here to include rallies, fund-raisers, protests, public meetings, coordinated letter-writing campaigns, petition drives, coordinated lobbying, and lawsuits brought by more than one person. This broad swath of community activity is best captured by the term "contentious politics," defined by McAdam et al. to mean "episodic, public, collective interaction among makers of claims and their objects when (a) at least one government is a claimant, an object of claims, or a party to the claims and (b) the claims would, if realized, affect the interests of at least one of the claimants" (2001, 5).

In the cases considered for this study, the contention was sustained over an episode of at least 102 days. The makers of claims were ordinary citizens in the community, who were often joined by local elected officials. The object of their claims was the state government administering the LLRW siting process. The claims themselves were opposition to the proposed location of an LLRW facility. These claims, when realized, affected both the claim makers and their object. Such a claim freed the candidate community from the environmental and economic consequences of hosting such a facility, and it thwarted the state from complying with its siting obligations under federal law and regional agreements.

For this analysis, I gathered data on contentious politics in each county by tracking all collective acts of public opposition to the LLRW facility in the local daily newspaper nearest to each site, from the day the candidate site was announced in the newspaper to the day the siting process effectively ended for each site. Admittedly, local newspaper coverage of any given issue can vary considerably across the United States, according to the resources the paper has at its disposal, the political whim of the editors and reporters, and the other local issues in the news at the time. Nevertheless, local daily papers are the best and only source from which to cull the day-to-day contentious politics that are simply not covered by larger media outlets. I cross-checked each local source with the nearest regional newspaper to ensure that a local paper was not grossly underreporting the events surrounding the LLRW issue and found that the regional sources did not report on a single event that was not covered by the local daily papers. In each case, the local daily papers gave the LLRW issue front-page status alongside the major national and international news stories of the day, which included the Iran-Contra scandal, Exxon *Valdez* oil spill, fall of the Berlin Wall, Tiananmen Square protests, and Oklahoma City bombing.

Figure 4.1 displays the number of collective acts of public opposition within a 100-day period after the first event in each of the candidate counties. This measure controls for the different durations of siting processes across these cases

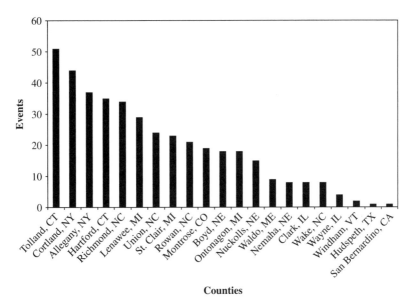

Counties

Figure 4-1. Collective Events of Public Opposition against LLRW Sites in the 100 Days Following the First Event in Each Candidate County

by confining the event count to 100 days. It also helps control for different rates of initial organization across the candidate counties by beginning the event count with the first collective event of public opposition in each case.

The Facility Siting Hypothesis: Limitations in the Quantitative Analysis

We saw in the last chapter that demographic profiling during the site selection did not enable siting authorities to avoid an opposition frame, but could key demographic factors explain the variations in levels of *active opposition* indicated in Figure 4.1? Quantitative analysis included in the appendix to this book—exploring the impact of the political leanings of elected officials on environmental issues, the ratio of the population defined as "rural," the median household income, and the educational attainment of the candidate county populations on the number of collective acts of public opposition in these communities—does not square well with the demographic profiling approach. The ratio of the county population defined as rural and educational attainment are significantly related to active opposition—but not in the expected direction. The more rural and less educated the county population, the more actively opposed it was during the LLRW siting process. These results do not match the expectations of the siting industry recommendations. Median household income does display a positive relationship with activism, as the industry experts expected, but the standard error is close to zero, and the political leanings of elected officials on environmental issues are insignificant. Overall, the model leaves much of the variation unexplained.

If the LLRW siting agencies employed the industry recommendations to help identify counties that would not actively oppose the proposed facility, they must have been disappointed with the results. As will become clear later in the chapter, many counties with demographic characteristics thought to predict inactivity mounted powerful opposition campaigns to the proposed LLRW sites. And some counties with characteristics thought to predict resistance actually mounted relatively little active opposition.

Classic Social Movement Hypothesis:
Limitations in the Quantitative Analysis

If social movement scholars had been given the job of consulting on the location of LLRW facilities, they would not have done much better than the industry professionals. The classic social movement agenda puts forth three explanatory concepts for mobilization, which would seem to inform our understanding of the nature of siting disputes in these cases. Like the industry professionals, social movement scholars recognize a role for the local political context with the concept of "political opportunity structure," which includes political access for movement participants and the presence of allies in power (McAdam 1996; Tarrow 1998). Social movement scholars also consider mobilizing structures and collective action frames as factors facilitating mobilization. Mobilizing structures are "embedded social networks and connective structures" (Tarrow 1998, 23) that link activists to "other groups" and provide "external support" (McCarthy and Zald 1987, 16). The concept of mobilizing structure is closely related to the concept of social capital put forth in Putnam's book *Bowling Alone: The Collapse and Revival of American Community* (2000). Civic organizations like clubs, churches, and fraternal groups are the essential elements used to evaluate the presence, absence, and amount of social connectedness in a community. To measure this concept,

I compiled a list of local civic organizations for each county using phone book listings, the identification of local chapters for each organization listed in the appendix of Putnam (2000), and organizations covered in the local daily newspapers during the siting process.

Collective action frames describe the strategic process of ascribing meaning to movement goals and activity (Snow et al. 1986). An "injustice frame" offers the best demonstration of the framing (Gamson 1992; McAdam 1999). When activists frame a contentious issue as an injustice, this creates a powerful political perception among individuals that helps mobilize support for the movement. Many environmental justice scholars use the "injustice frame" to explain mobilization against community environmental harms (Aronson 1997; McGurty 1995; Novotny 1995). To assess the pervasiveness of an injustice frame, I identified letters to the editor in the local papers expressing a message either that the state government or siting agencies were motivated by political rather than technical considerations or that the state government was not representing the residents' interests.

Table A.2 in the appendix to this book displays the results of Poisson regression evaluating the impact of political opportunity structures, mobilizing structures, and collective action frames on acts of opposition in counties facing an LLRW facility proposal. Only the measures of political opportunity and the prevalence of an injustice frame are significant. Political opportunity is not significant in the expected direction.

Overall, this model explains a modest amount of active opposition, yet the prevalence of an injustice frame is significant and positively related to active opposition. This result supports evidence gathered from case studies on facility siting that highlight the role of an injustice frame in mobilizing community opposition (Aronson 1997; McGurty 1995; Novotny 1995). This is the only independent variable in the classic social movement agenda model that measures a phenomenon that is actively created during the siting process. Political opportunity structures and mobilizing structures are preexisting conditions in the candidate counties. The significance of the framing variable, which measures a dynamic process during the episode of contention, should shift research and analysis away from preexisting factors to dynamic aspects of the process that influence mobilization. McAdam explains the importance of framing relative to the other key factors of the social movement model this way: "While important, expanding political opportunities and indigenous organizations do not, in any simple sense, produce a social movement. In the absence of one other crucial process these two factors remain necessary, but insufficient, causes of insurgency. ... Mediating between opportunity and action are people and the subjective meanings they attach to their situations" (1999, 48). The qualitative analysis of carefully selected cases in the rest of this chapter is better suited to explore such dynamic interactions associated with mobilization.

A QUALITATIVE APPROACH: UNEXPECTED DIFFERENCES BETWEEN COUNTIES

The count data on the collective acts of public opposition (Table 4.1) can help identify a carefully paired comparison of one county that displayed frequent

County	Events
Tolland, CT	51
Cortland, NY	44
Allegany, NY	37
Hartford, CT	35
Richmond, NC	34
Lenawee, MI	29
Union, NC	24
St. Clair, MI	23
Rowan, NC	21
Montrose, CO	19
Boyd, NE	18
Ontonagon, MI	18
Nuckolls, NE	15
Waldo, ME	9
Nemaha, NE	8
Clark, IL	8
Wake, NC	8
Wayne, IL	4
Windham, VT	2
Hudspeth, TX	1
San Bernardino, CA	1

Table 4-1. Collective Events of Public Opposition against LLRW Sites in the 100 Days Following the First Event in Each Candidate County

acts of collective public opposition to the LLRW facility, and another case that was not very active at all in this regard.

The two counties that are the best candidates for this paired comparison are Richmond and Wake Counties, both in North Carolina. Richmond County is in the upper quartile of active communities, while Wake is in the lower quartile. No other state has two counties with such different levels of active opposition. Richmond County generated more than four times the number of collective events of public opposition than Wake County during the 100 days after the first opposition event in each county. These two counties are also among the five counties that endured a siting process of more than 1,000 days. This similarity helps control for the effect of duration on collective acts of public opposition. The letters to the editor in the local daily papers of both of these counties expressed overwhelming opposition to the LLRW site (93 percent of letters to the editor in Richmond County and 80 percent in Wake). Both county governments also formally opposed the proposed LLRW site.

The qualitative analysis that follows on the successful mobilization of opposition in Richmond County and the frustrated efforts of activists opposed to the Wake County site is based on semistructured qualitative interviews and local newspaper accounts of the siting process. In 2003, I conducted 33 interviews with activists, local and state government officials, newspaper reporters, members of the clergy, and businesspeople in each community, many of whose comments appear below. I also identified and reproduced every newspaper article on the LLRW issue in each county's daily paper for the duration of the siting processes.

Erroneous Assessments by the Industry Consultants

The consultants hired to evaluate the likely community responses in North Carolina counties employed demographic and political variables similar to those considered in the analysis above and concluded that citizens in Richmond County would be less actively opposed to the LLRW site than citizens in Wake County. The result, however, was just the opposite.

Robert E. Leak, who was hired by Chem-Nuclear Systems, the firm charged with siting an LLRW facility in North Carolina, advised his client that Richmond County would be "among the less difficult of the potential sites" in which to gain community acceptance, or at least acquiescence. The consultant correctly identified Richmond as a poor county run by a conservative, single-party machine. Richmond County had the highest unemployment rate and the lowest median household income of the counties under consideration. Leak reported to Chem-Nuclear optimistically that the Richmond economy was "flat" and "two plants have closed and two others are rumored to be about to close" (Epley 1989, DD06352, D009220). He assumed that local politicians could be persuaded to back the LLRW facility, and that a submissive community would follow the lead of its local political establishment. Other consultants noted that the county had no history of environmental activism— or activism of any kind.

In contrast, Leak viewed the candidate site on the Wake–Chatham County border (which was later redrawn to fit entirely within Wake County) as problematic, warning that this site "may attract the most attention and debate" and "pressure is likely to be intense" from the public at large opposed to the site. His recommendations noted that Wake County was "one of the state's most urban, prosperous and populated regions," and numerous environmental groups would "be able to generate a great deal of opposition among neighborhood groups in Raleigh … and in adjoining counties" (Epley 1989, DD06167). Perhaps also because of these same characteristics, some environmentalists lobbying the North Carolina legislature during the earliest phases of the siting process were convinced that although the Wake site was on the short list of candidate sites, it was a "red herring." One environmentalist, Lisa Finaldi, said she thought Richmond was the preferred site. According to another, Janet Zeller, "The Wake site was never going to be the site. I was shown a note in which high state officials had said that Wake/Chatham sites would not be used."

The consultants mischaracterized these two communities. Table 4.2 shows that residents in Richmond County mounted many more collective events of public opposition to the LLRW site proposal than those in the Wake County area. Throughout the course of the siting struggle, Richmond residents carried out nearly four times the events organized by Wake area residents, even though the siting process in Wake went on two years longer than the process in Richmond. Richmond averaged one collective act of public opposition every 11.98 days over the course of the struggle, whereas Wake County averaged just one event every 68.62 days.

County	Events in 100 days after the first event	Total events	Siting process duration in days	Duration/events
Richmond	34	123	1,473	11.98
Wake	8	34	2,197	68.62

Table 4-2. Collective Acts of Public Opposition to a Proposed LLRW Facility in Richmond and Wake Counties
Note: Events include all collective public acts of opposition concerning the site, even those that occurred in neighboring counties

All of the activists and elected officials I interviewed across both counties acknowledged the fact that Richmond County residents opposed the LLRW facility proposal more actively than the Wake County area residents.

Once it was clear that Richmond was mounting much more active opposition than Wake, Leak dramatically revised his assessment of the two counties. He reported to Chem-Nuclear in August 1990 that there was "no visible support anywhere" in Richmond County for the LLRW site. Conversely, in the same report, Leak concluded that "Wake County is simply not concerned except for a few people in the immediate site area" (Epley 1989, D009219–D009220). Robert Leak told me in an interview that after the candidate sites were announced, he was actually threatened by an elected official representing Richmond County. The official reportedly approached Leak while he was dining with his wife in a Raleigh restaurant and warned him not to return to Richmond County out of concern for his safety. After that, Robert Leak vowed he would never again get involved in a project of this type and changed his firm's motto from "We'll do economic development anytime, anywhere for anybody" to "We'll do economic development anytime, anywhere for anybody as long as it's fun."

The Weaknesses of the Classic Social Movement Agenda

Leak's misreading of these two counties is instructive for social movement scholars. His judgment suffered from a number of blind spots that are relevant to the way social movement scholars have attempted to identify factors of mobilization. Leak's socioeconomic profile followed in the tradition of the Cerrell report (1984) by using census data and surveys to measure and aggregate individual characteristics. To this he added descriptions of preexisting structural factors in the community. But he failed to imagine, predict, or fully consider the ways patterns of relationships would emerge in these communities between and among individuals—and the political dynamics those relationships would generate. Similarly, critics of the classic social movement agenda, including founding scholars like McAdam and colleagues, argue that this scholarly approach also relies too heavily on static "objective structural factors" to explain mobilization (2001, 43). In these two counties, consideration of such structural factors would wrongly predict more active opposition in Wake than Richmond.

Political Opportunity Structure

"Political opportunity structure" was the most widely employed and troublesome concept to emerge from this literature. In a 1994 synthesis of the literature, Tarrow wrote that "people join in social movements in response to political opportunities." He argued that "political opportunity structure" explained the "when" of social movement mobilization. Tarrow made it clear that the concept emphasized "resources external to the group" in the political environment that "either encourage or discourage people from using collective action" (1994, 17–18). Resources thought to influence mobilization include institutional access for movement participants, rifts among elites, the presence of allies of the movement in power, and a decline in repression (McAdam 1996; Tarrow 1998).

At the time North Carolina announced the LLRW candidate sites, Wake County enjoyed a wealth of political opportunities that would facilitate active opposition to the LLRW site, while Richmond County seemed to lack any such opportunities. The Democrats were the majority party on each county's board of commissioners, and each county's delegate in the U.S. House of Representatives was a Democrat. However, the character of the Democratic Party in these two counties was very different.

The longtime conservative Democratic Party in Wake County was under siege in the 1980s from tens of thousands of newly transplanted northerners taking professional positions in the Research Triangle. Institutional access, which had been historically limited to what one activist, Liz Cullington, described as "big money, quite conservative Raleighites," was now opened to the challenge of "socially liberal Republicans and Democrats from New York." This influx of professional, affluent transplants created instability in the political system. The consultants for Chem-Nuclear reported in 1989 that in Wake County, "there is no single political power" (Epley 1989, DD06167).

One county official recalled that the Wake County Commission was "sort of divided" on the LLRW issue as a result of the different perspectives of "new people" and those who had lived there all their lives. This new and unstable political environment offered environmental activists some allies in power. Each activist I interviewed in the Wake County area identified State Representative Joe Hackney and State Senator Eleanor Kinnaird as strong allies in environmental causes. The League of Conservation Voters (LCV) score for the Wake-area member of the U.S. House of Representatives was 90 percent, which indicates a high level of political support on environmental issues. Finally, none of the activists mentioned a fear of repression from local or state law enforcement.

In contrast, Richmond County was dominated by a conservative Democratic political machine that guarded institutional access, ensured political stability, was not sympathetic to environmental concerns, and effectively employed a threat of repression. In 1989, the Democratic sheriff R. W. Goodman had been the political boss of Richmond County for nearly 50 years. In addition to his position as sheriff, Goodman owned a department store and a textile mill and sat on the board of directors of two Richmond County banks. Sheriff Goodman carefully guarded institutional access. According to a political reporter for the local daily paper, whom I also interviewed in 2003, Goodman "handpicked" all local and state officials representing Richmond County. Goodman's Democratic Party was stable and solidly conservative. One elected official at the municipal level summarized the one-party system this way: "If you're not a Democrat in this county you can forget it. You can't be a dog catcher if you're Republican here." Former state senator Richard Conder told me that Richmond County "has been a strong Democratic County since day one." He also characterized the county as "very conservative," declaring that "we don't have much liberalism in Richmond County." The conservative party machine did not offer environmental activists any significant allies in power. Both newspaper editor Glen Sumpter and local official Abbie Covington told me that the environment had never been a big issue for the area. The LCV scorecard for the Richmond representative in the U.S. House of Representatives was just 50 percent, very low for a Democrat in federal office.

Finally, the Democratic Party machine in Richmond effectively raised the specter of repression. One LLRW activist confided that he worried about "ending up in the river." Another explained that "you don't even think about civil disobedience—they're going to whoop your butt; they'll open a can of whoop ass." Three outside environmental organizers told me that many residents complained of "intimidation" from state and local law enforcement personnel. These same organizers said they were monitored and followed by law enforcement officials.

Mobilizing Structures
In the quantitative analysis earlier in the chapter, Richmond County exhibited a greater density of civic organizations than Wake County. Richmond had one civic organization for every 291 people, while Wake had one civic organization for every 381. Relative to the rest of the candidate counties, however, Richmond and Wake rank closely on this measure of mobilizing structures, as Richmond ranks 12th and Wake ranks 14th among the 21 counties. More important, Wake County hosted at least 16 environmental organizations, whereas Richmond did not host any such groups. Consultants working for Chem-Nuclear found that Richmond County had "no history of environmental controversies or environmental group activism" and argued that the "lack of environmental activism within the county" could make siting the facility "less difficult." These same consultants found that environmental groups in Wake County had "successfully thwarted" the siting of hazardous waste incinerators and depositories (Epley 1989, DD06352, DD06177). Practitioners of the classic social movement agenda would argue that these existing environmental mobilizing structures with experience organizing active opposition would lead to greater opposition to the LLRW site in Wake County. Yet Richmond, the county with no existing environmental mobilizing structures, organized far more collective acts of public opposition.

Collective Action Frames
If there is an element of the classic social movement agenda that could indicate greater active opposition in Richmond than in Wake, it is the prevalence of the

County	Percentage of letters to the editor with an injustice frame	Percentage of days during siting process when injustice frame appeared on editorial page
Average across 21 candidate counties	40	6
Richmond County	36	4
Wake County	26	1

Table 4-3. Use of the Injustice Frame in Richmond and Wake Counties

injustice frame on the editorial pages of the local papers. This is the frame scholars of the environmental justice movement have suggested influences community mobilization. Residents writing letters to the editor in Richmond County did apply the injustice frame more than those writing in Wake County (Table 4.3).

However, both of these counties applied the injustice frame with below-average frequency on the editorial page of the local daily paper relative to the average among other candidate counties. Neither of these counties ranks extremely high or low relative to the other candidate counties in its use of the injustice frame. When letters with an injustice frame are measured as a percentage of all letters to the editor, Richmond ranks 12th, with 36 percent, and Wake County ranks 16th, with 26 percent. When the letters with an injustice frame are calculated as a percentage of total days during the siting process, Richmond ranks 7th and Wake ranks 13th.

MOVING FROM THE STATIC TO THE DYNAMIC

The quantitative analysis included in the appendix to this book found a significant and positive relationship between the prevalence of an injustice frame and active opposition. On closer examination, this is also the only aspect of the classic social movement agenda that could help explain the higher number of collective acts of public opposition in Richmond County. As measured above, the collective action frame variable is the only aspect of the classic social movement agenda that is dynamically created during the episode of contention. The variables operationalizing political opportunity structures and mobilizing structures measure static preexisting conditions before the episode of contention began. The preexisting structural factors present in the political environments of Richmond and Wake were not irrelevant, but they were by no means determinant factors.

Fortunately, the "classic social movement agenda" was never this simple, and theoretical development has not halted. First, while Tarrow argued that "people join in social movements in response to political opportunities," he also argued that collective action may "create" political opportunities. He aptly identified this two-part process as "seizing and making opportunities," indicating an active process (1994, 17, 81). Seven years later, McAdam et al. (2001) revamped the classic social movement agenda to focus on the "dynamics of contention." This was a major theoretical step away from preexisting structural factors to an exploration of social mechanisms working during the episode of contention. The dynamics of contention approach attempts "to identify the dynamic mechanisms that bring these variables into relation with one another and with other significant actors" (2001, 43).

Activists are indeed "active" in shaping and creating the political environment in which they develop. On the whole, the preexisting objective measures identified by facility siting consultants and the classic social movement agenda indicate that Wake County should have been more active than Richmond. The key explanatory difference lies in the way activists in each community utilized their existing resources. I take up the challenge laid out by McAdam et al. to "identify crucial mechanisms and robust processes that produce the distinctive features" of the episodes of contention in question (2001, 84–85). Elster tells us that mechanisms are "frequently occurring and easily recognizable causal patterns" (1998, 45). Each of the mechanisms I identify in these two cases qualify as "frequently occurring and easily recognizable"; they have each been spotted, tagged, and defined over a diverse range of contentious politics by McAdam and his colleagues. I find that the interesting differences between Richmond and Wake County lie in the setting, sequence, and combination of the following social mechanisms as defined by McAdam et al.:

Social Appropriation: the active appropriation of social sites for mobilization.

Identity Shift: alteration in shared definitions of a boundary between two political actors and of relations across that boundary.

Attribution of Threat and Opportunity: the challengers' perception of their political context. (2001, 43–44, 162)

In Richmond County, activist leaders overcame obstacles to mobilization in their setting, such as geographic and racial divisions, by appropriating powerful social sites and forming a group identity that unified previously disparate county groups. Once this was accomplished, Richmond County activists and the county at large began to perceive political opportunities in their initially hostile local political context. In Wake County, activists were unable to achieve robust active opposition because they failed to appropriate any countywide social sites, and they could not link the various area communities to affect a shift toward countywide identity. Although Wake County enjoyed some preexisting political opportunities, local activists did not perceive their local government context as one of opportunity. This conspicuous absence of social appropriation, identity shift, and attribution of opportunity ensured that Wake activists would fail to overcome preexisting structural obstacles to mobilization or take advantage of preexisting structural opportunities.

STARTING FROM SCRATCH: CREATING POLITICAL OPPORTUNITIES IN RICHMOND

Richmond County resident Bobby Quick told me that on the night of November 8, 1989, he said to his wife, Beverly, "We'd better get a good night sleep now, because it may be the last one we get for a good long while." He had just read the first article in the local paper on the LLRW site and thought, "Well, that's just a couple of miles down the road there, and we might need to consider fighting this." In these first days, Quick and the original leaders of the opposition in Richmond County were virtually alone in their outspoken opposition to the proposed LLRW site.

Most Richmond County residents did not entertain the possibility of challenging the LLRW facility proposal. On the most fundamental level, one

Richmond activist told me that most people were intimidated and would say things like "I would love to be able to take part in these meetings, but I can't because of my job." Lisa Finaldi, an outside organizer who was involved during the first month of the siting struggle, said, "There was also a lot of talk that when they had tried to organize things in the past on their own, there was a lot of intimidation by the sheriff." Another activist, Janet Zeller, told me that initially "people were afraid to speak out against the dump, because there was the belief in the county that this wouldn't be happening if the sheriff was not in favor of it." As one future opposition leader, David Ariail, put it, "Many believed the sun didn't shine and the rain didn't fall unless the sheriff allowed it." Bobby Quick described a situation in which a woman was signing a petition opposing the LLRW facility when her husband approached and said, "Don't sign that thing; the same people that's been looking after us all these years will look after us again." Even future leaders of the opposition said that they initially hesitated to act because they did not think that they could do anything about it. One told me, "I didn't think that anyone was going to try to resist having the site here." Ariail, who became the first chairperson of the LLRW opposition group, said, "Most people in the community were confused or apathetic. Some felt it was a done deal ... of course, people felt the sheriff knew the real story, and if he wanted it here, there was nothing anyone could do about it."

Social Appropriation

The first collective act of opposition in Richmond was a petition campaign led by Bobby and Beverly Quick. Most people were comfortable signing the petition even if they were not yet willing to fully commit to the opposition. As one signatory told Bobby Quick, "I'll sign it ... but you'll never be able to get people to pull together on this." The signatory was wrong. The fledgling opposition group was able to unite the county by appropriating the most active social site in the county: the Quicks launched their opposition campaign by gathering signatures for the petition at the Richmond County high school's football games.

Desegregation and countywide consolidation of the school district in the 1970s had created a powerhouse football team. In 1989, the Richmond Raiders were on the way to their second undefeated season as state champions. When the LLRW facility was announced in November, the editor of the local paper wrote that "this doesn't seem like a year with a lot of things to be thankful about." He recounted the damage of Hurricane Hugo and the threat of the proposed LLRW site. Then he focused on the one thing all Richmond County residents *could* give thanks for: "The football team is winning. In fact, the Raiders haven't lost since the Middle Ages" (Sumpter 1989b). Several interviewees made me aware of the football team before we even discussed the LLRW issue. One activist explained, "In a small community, you either have something or you don't, and you get excited about it." Richmond County had its state champions, and "more people came together for those football games than anything else." The state championship forged a county identity that transcended historic divisions among different Richmond municipalities and eased tensions between black and white residents.

The petition campaign at the football games brought committed activists together, informed the public of the issue, and created a record of public opposition. Several members of the Quicks' high school class of 1967 became reacquainted at the football games that season. These people helped form a citizens group opposed to the LLRW facility. One 1967 alumnus told me, "My first 'real' conversation about the siting was probably with Bobby and Beverly Quick at a local football game. ... I really didn't know until that time that anyone was going to try to resist having the site here. So when I saw that they were, we got together and we started making plans."

This group joined other opponents in the county to form For Richmond County Environment (FORRCE) in late November 1989. When the first official meeting convened the Quicks' petition campaign had gathered 6,300 signatures, amounting to 14 percent of the county's population. The local newspaper covered the FORRCE meeting and took statements from group leaders. The petition was featured prominently in the story, and Beverly Quick referred to it as proof of widespread opposition. "First I was not sure how the people felt," she said. "Now I am thoroughly convinced that they are opposed" (Sumpter 1989a). When the petition drive was completed, FORRCE had gathered 26,756 signatures—more than 60 percent of the Richmond County population (AP 1990a).

Opponents of the facility actively framed the issue in terms of Richmond County identity, linking the football team to the struggle against the LLRW facility. One letter to the editor read: "Should this waste facility enter our county, it would become infamous as the 'pollution capitol [sic] of North Carolina' not famous for its 'Raider' spirit, textile industry, or agricultural heritage. Citizens of Richmond County, let us strive to keep 'Raider' country 'green'" (Thomas 1989).

Ruby W. Dunn wrote a letter in the paper encouraging people to support FORRCE, saying, "Stand up and be counted, be on the winning team, stick together and together we will be victorious" (Dunn 1990). The football team was a tangible and universal element of Richmond County's identity. So opponents of the LLRW facility could use it to stir the emotions of all Richmond residents. LLRW facility opponents marching in the annual Christmas parade carried a huge football scoreboard that read "Richmond County 10, Chem-Nuclear 0." The Richmond Raiders themselves marched in uniform during rallies throughout the siting struggle.

Identity Shift

Perhaps because of this appropriation of the most salient social site in the county, FORRCE was reportedly the first citizen group to effectively bridge both black and white communities. FORRCE was initiating a countywide identity shift that would bring previously separated elements of the population together to challenge the LLRW site. Activist Lisa Finaldi reported that prior to FORRCE, people of different races "would never go to a meeting together or sit down next to each other." She said this issue was "the first battle to overcome ... we needed to have one meeting where everyone was present rather than going all around town." Another activist claimed that "Sheriff Goodman had more or less carved up that county into black and white communities. ... He had them chopped up every which way to control them." She argued that FORRCE was "the first time in the history of the county that the black and white people worked together" and characterized the group as "an almost picture perfect joining of black and white together in a real unified campaign." Activists noted that FORRCE also enjoyed the support of both the predominantly black and predominantly white churches in the area.

The FORRCE membership also joined residents from the two major county municipalities of Hamlet in the south and Rockingham in the north. At the same time that a site in the south of the county was a candidate for the LLRW facility, a site in the north of the county was under consideration for a hazardous waste incinerator. FORRCE took on both issues. As one activist recalled, "The first meeting brought together people from the northern part and people from the southern part" of the county. "The northern people were all enraged about the incinerator site, and the southern people were enraged about the radioactive waste dump." David Ariail believed that "fighting on two fronts" made FORRCE "a stronger and bigger operation." Ariail was the director of the community theater in the county seat of Rockingham, while other leaders like the Quicks were from Hamlet. Bobby Quick believed that the incinerator issue generated "a lot more followers" for FORRCE. Before the incinerator was proposed, "people would say, 'Well, you know, if I lived down in that end of the county, I'd be more interested,'" said Quick, "but when they popped that one up at the other end of the county, that really helped us."

Attribution of Political Opportunity

The petition drive and organizational meetings were initial steps in the FORRCE effort to capture the attention and support of the local elected officials. Whereas many residents saw the Richmond County political machine as a juggernaut, FORRCE activists saw a vulnerability they could exploit. FORRCE leaders perceived political opportunities rather than threats. Ariail said of the struggle, "The one thing I learned was that politicians are very vulnerable. They tend to think that they run the show, but in reality they were on the run." Bobby Quick explained that the FORRCE strategy was "to just pull them [the politicians] over to your side and make it popular for them to be against it. ... Eventually we got all the [county] commissioners to be on our side."

A one-party machine does not exist apart from its host community. To some degree, the system depends on legitimacy granted by the larger public. In Richmond County, as elsewhere, Ariail pointed out, "Most politicians want to be reelected." Newspaper editor Glen Sumpter noted that Sheriff Goodman "did not go against the tide" of public opinion in the county. Instead, it seemed that the sheriff was trying to distance himself from the issue by not taking a position on it, possibly so as not to alienate the voters and harm his chances of reelection. Bobby Quick said that he was able to "convince a lot of others that instead of going after politicians and trying to get a vendetta against them, or trying to get rid of them or anything like that while the thing's going on, you want to get them on your side, or give the appearance of being on your side, whether they are or not . . . at least get them on the record [as being opposed to the site], and then once they're on the record, it's harder for them to go back."

Two local dynamics helped support this strategy. First, Richmond County politics take place via face-to-face relations on a daily basis. As one elected official, Abbie Covington, explained, "In a county as small as ours, there's obviously a lot of personal as well as professional involvement in politics, because we run into the people we work for in the grocery store." State Senator Richard Conder described his constituency contact similarly: "You know, in a county that small . . . I know just about everybody in the county. . . . They catch me in church on Sunday, they catch me on the street." One movement leader said that when you go to church with your county commissioners, "they have to come across with at least part of what you were asking them to do." The personal contact of small-town politics also cultivates gossip and rumors. Ariail noted that elected officials "were worried about what we knew and how much we knew." He remarked, "I never let on what we knew." One activist told me he got a call from a local politician's wife asking him to "please protect my husband," which the activist interpreted as a plea to make sure the husband would not lose his position.

The second dynamic that helped the opposition was the uncertainty of the sheriff's position. Arial noted that "his blessing determined the outcome of many elections. But he was very quiet about the siting process." Robert E. Leak's report to Chem-Nuclear claimed that the sheriff did not oppose the LLRW facility, but as Leak told me later, "He sure as hell didn't stand up for it either." Internal documents show that Sheriff Goodman indicated to consultants that "the matter was not his problem" (Epley 1989, DD06097). The sheriff's silence on the issue gave the activists a political space in which to pressure their other elected officials. The political machine was not leading on this issue, which, Ariail explained, led the opponents to develop an attitude that "the people were the authority."

Newspaper coverage of the first FORRCE meeting reported on the petition as a record of public opposition and also noted that the main concern of this first meeting was "the lack of public opposition by elected and appointed officials." David Ariail was elected chairman of FORRCE that night, and in his first public statement, he issued a challenge to the local government: "People want to know the position of our elected officials on this. It's fair and honest to demand our elected officials to come forth in such an important time. They have a responsibility to say to the County where they stand and what they can do about it" (Holland 1989a).

The first major event that FORRCE organized was an attempt to bring the public and the elected officials together in a forum where the public could directly confront their representatives. The event was billed as a public hearing, a familiar format in the county courthouse, where government officials would respond directly to public comments. However, unlike most public hearings, this was not organized by a government body. FORRCE organized the hearing and was therefore able to preside over the question-and-answer session.

The elected officials felt compelled to attend, and the general public felt comfortable participating. FORRCE members were able to shape the debate, with speeches at the beginning of the meeting. Then local officials were each pressed to address the crowd. The local newspaper reported: "The crowd overwhelmingly expressed its opposition. ... However, political figures on the panel may not have been prepared for the public suspicion that followed ... several persons challenged panel members and other elected officials for having shown minimal leadership during the organizational efforts in the county to oppose the placement of a disposal facility here" (Holland 1989c).

Bob Maloney, a Hamlet city councilman and candidate for the County Commission, leveled charges against other municipal governments and named members of the County Commission and each state senator and representative, explicitly challenging them to go on record against the LLRW facility. Once Maloney started this line of questioning, he was joined by a long line of audience members. The FORRCE members running the hearing let the public continue to vent their frustrations, as the elected officials remained trapped on stage, forced to respond.

Bobby Quick described the event this way: "We had these politicians up there, and people started asking them questions, and that was the same time that they realized, 'Hey, we'd better go over their way and look like we're interested.' And from then on they did. I believe that we eventually won them over to our side." Ariail also saw a transformation in the elected officials: "It was like everyone got new marching orders." He observed that local officials began to actively oppose the dump "to convince citizens that Chem-Nuke was never invited to Richmond County." Numerous letters to the editor criticized the local officials, with warnings like this one: "People of Richmond County are waking up to our 'do-nothing' elected officials" (Shankle 1990). There were even calls for the ouster of Sheriff Goodman (*Richmond County Daily Journal* 1990; Hollingsworth 1990). One letter called for a new sheriff and declared: "It's about time we realize the need for strong leadership in our county. We don't need a 'free' turkey for Christmas. Take a good look at the condition of our county and you might see the high price we've paid for a 'free' turkey" (Swicegood 1990). The sheriff eventually expressed his opposition to the facility while he spoke on the issue at a FORRCE meeting.

FORRCE had created political opportunities by forcing the county's elected representatives to take a stand on the LLRW issue. As one county commissioner recalled, "They just came in droves to the commission meeting and blamed us for it and said, 'Listen, you've got to do something about it.' And we were besieged by people." Elected officials at all levels of government responded to the public challenge. State Senator Conder vowed "to fight efforts to locate the dump in Richmond County" and claimed he was in regular attendance at FORRCE meetings and events (*Richmond County Daily Journal* 1989b). State Representative Don Dawkins wrote a letter to the editor that read: "I have done everything I could do, publicly and privately, to oppose the siting of a low-level radioactive waste dump in Richmond County. ... Elected officials and concerned citizens can keep the eight-state low-level hazardous waste dump out of Richmond County" (Dawkins 1989).

Dawkins's wife, Pat, helped coordinate fund-raising for FORRCE. Prentice Taylor, chairman of the County Board of Commissioners, pledged to "do everything we can legally, morally and spiritually to fight this thing" (Sumpter 1990). The Board of Commissioners opposed the site and unanimously appropriated funds with which to hire a zoologist, geologist, lawyer, and other experts to gather evidence discrediting the site selection. The county manager, whom Chem-Nuclear consultants initially characterized as a possible supporter of the facility, became an outspoken critic of the siting process in numerous government meetings in Raleigh. The first chairman of the Richmond County

Site Designation Review Committee (SDRC) was the former North Carolina secretary of natural resources, another prominent individual identified by the consultants as a possible supporter of the facility. At the first SDRC meeting, he declared that the committee's purpose was "to prove this site unsuitable" (*Richmond County Daily Journal* 1989a). The county health department issued a report claiming that the site was not geologically sound (*Richmond County Daily Journal* 1989c). All the county's municipalities went on record as opposing the site proposal.

Every official seemed to feel that he or she had to go on record against the LLRW facility. One official reportedly went to Raleigh to give a speech against the facility and nervously fumbled during his presentation. In his confusion, he reached out for the one touchstone close to all Richmond residents: "I tell you what, we're good people, and I tell you what, you get any football team in the state and bring 'em here, and we'll beat the pants off of 'em."

Ariail said at this point, "I realized we had won." Although the Richmond site was still under serious consideration, "the suspicion that veiled our community from the beginning of an insider sellout was over. Now the county politicians were committing to spend whatever it took to legally fight the dump." County commissioners were even turning to Ariail during meetings and asking him what he thought about various decisions. Commission members met with FORRCE and the county's legal team to plan strategy (MacCallum 1990; Sumpter 1990). Another FORRCE member commented, "I think for a long time in this county, up until that point, the politicians felt like you could tell people anything you wanted and they'd believe you. But they found out a little differently. ... They followed the people, just like a good politician would."

One statewide activist, Lou Zeller, said the politicians in Richmond were most influenced by the unprecedented countywide unity FORRCE had cultivated. "I think to cross those boundaries and unify the people, black and white as well as across the county, in a way that threatened [Sheriff Goodman's] control ... was one of the things that started to turn the leadership of the county around. They realized that if they didn't move and act, they would be in trouble. ... Once the community had a belief that they could do it, a bridge was crossed. Then the officials were really just an apparition of power, and the savvy politicians said, "I gotta get in front of this."

Successful Mobilization

A little over two months after the site proposal was first announced, Richmond County hosted speaker Brian Young, the southern regional director of Greenpeace. More than 1,000 residents crowded into a junior high school auditorium to hear him speak. Every local official of significance was on stage with Young, expressing opposition to the LLRW facility. This panel included Sheriff Goodman, Senator Conder, Representative Dawkins, several county commissioners and municipal officials, and Mike Ezzell of the United Carolina Bank, representing the business community. One week later, the sheriff's deputies turned on their sirens and led a 2.5-mile motorcade of Richmond county residents to Raleigh. After circling Raleigh, more than 1,200 residents disembarked from cars and buses and marched on the governor's mansion. Local officials accompanied FORRCE members to present the governor with a petition opposing the LLRW facility.

The month before FORRCE organized the countywide motorcade to Raleigh and the march on the governor's mansion, hundreds of Richmond County residents drove to Chapel Hill to watch the Raiders defeat A. C. Reynolds High for the state championship. In the lead-up to this event, the Raiders' Booster Club helped arrange the trip and encouraged fans to celebrate "Raider Day" by displaying their green and gold Raider flags (Brigman 1989). FORRCE took a strikingly similar approach the next month, encouraging Richmond County residents to celebrate "Commitment Day" by committing to take action against the LLRW facility and fly their American flags at half-mast to "mourn the death of our county to nuclear waste" (FORRCE 1990).

FORRCE appropriated the football team as a starting point and progressed to constructing messages about the multigenerational heritage of families in Richmond County. Lisa Finaldi described the developing opposition as more of a "property rights" movement than an environmental movement. She said the opposition "was about somebody coming in here and telling [the landowners] what they had to do." Many letters to the editor outlined family lineages in the county that stretched back to the 18th century. A letter published in the December 12, 1989, *Richmond County Daily Journal* exemplified this message: "My relatives have lived in Richmond County since 1768. They farmed the land, worked at other occupations, paid taxes and obeyed the laws. They served in the military and some lost their lives in battle. Now the state wishes to repay this good citizenship by proposing the placing of a low-level radioactive waste 'dump.' … This dump will convert Richmond County from being a nice place to live to a future cemetery" (Gibson 1989).

While this message drew on a shared familial heritage in Richmond County, it also hinted at an outside force, "the state," invading the county to ultimately destroy it. This theme was more explicit in another letter to the editor published on March 11, 1990: "For a group of faceless people to designate us to accommodate such sites is truly sickening. Do they not care that these hazards will forever change our way of life here? Everything that people have worked for all of their lives could and probably will be essentially worthless" (Hollingsworth 1990).

The message was meant to get people to fight the siting process. The following letter used the analogy of an individual defending his hard-earned income with a pistol against potential thieves: "Those who knew 'Pop' will remember the pistol he carried in plain view whenever he carried his moneybag. As a boy, I distinctly remember his answer to me when I asked 'why the pistol.' 'Bubba,' he said, 'your mama and I have busted our rumps all our life to get the little we have today, and I'll be damned if I'm gonna let somebody take it from us, if I can help it.'" The writer continued: "That's why I am now an active member of FORRCE … I'm taking a stand. You see—it's my home. … How about you—Is this your home or are you just living here? Become active in FORRCE friends, 'cause we're gonna win this battle if we can help it" (Nettles 1990).

A large amount of social pressure was building that encouraged citizens to be active in opposing the LLRW facility. Ariail remarked that after the motorcade event and the march on the governor's mansion, "the question within Richmond County became 'Why didn't you go to Raleigh?'"

FORRCE successfully campaigned for a regular column in the local daily paper and a regular voice on the local radio station. As the editor of the paper, Glen Sumpter, told me, "FORRCE had gotten to be a fairly large local organization, and we decided we would give them a voice." As the struggle progressed, FORRCE appropriated the local government. Several FORRCE members sat on the county's Site Designation Review Committee to oversee the siting process. The city of Hamlet provided a permanent meeting place in the senior citizens center. Local officials attended FORRCE meetings, and officials and activists consulted together on strategy.

The episode of contention did not end until Richmond County was removed from the list of candidate sites in December 1993. But the activists had overcome the inertia of a one-party conservative county and actually created a favorable political environment with rich social support networks. The struggle against the LLRW facility had become more than a struggle of FORRCE against Chem-Nuclear—it was a struggle of Richmond County against Chem-Nuclear.

Finally, Richmond County successfully expanded the struggle outside its borders and appropriated social and political space in neighboring counties as well as South Carolina, as the site was on that state's border. A coalition of nearby counties lent support through the Community Action Network. And, as Raleigh resident George Miller told me, officials from South Carolina, including Senator Strom Thurmond, opposed the Richmond County site with so much vigor that most government officials agree that this fact helped remove the county from the list of candidate sites.

DEMOBILIZATION AND FRUSTRATION: THE RISING AND FALLING TIDE OF ACTIVISM IN THE WAKE-CHATHAM AREA

The consultants working for Chem-Nuclear reported that Wake County was "likely to be among the most vocal and active groups opposing the site. Public protests, media events and grassroots organizing are likely to be the group's tactics. ... They now know the political system very well" (Epley 1989, DD06178). Yet the proposed LLRW site in Wake attracted very few collective acts of public opposition relative to other candidate counties. The consultants based their analysis on Wake's growing economy, high median household income, and history of environmental activism. However, interviews with activists and community leaders in the Wake County area reveal several preexisting obstacles to active opposition. Some of these obstacles, such as municipal and racial divisions, were similar to those that activists in Richmond County overcame. Other obstacles, such as the presence of a nuclear power plant and a highly professional and transitional population, were unique to the

Wake County area. Activists opposed to the proposed LLRW site in Wake County were frustrated by their inability to sustain active countywide opposition in the face of these obstacles. The social mechanisms that could have facilitated active opposition, such as social appropriation, identity shift, and the attribution of political opportunity, were conspicuously absent during the episode of contention.

Like activists in Richmond County, leaders of the Wake-area opposition were faced with municipal and geographic obstacles. The proposed LLRW site originally spanned the boundary between Wake and Chatham Counties. Eventually the site was redrawn so that it was fully contained within Wake County. Nevertheless, the affected populations were located in both counties. "In the South," said Mac Legerton, a clergyman and statewide social justice activist, "a county may as well be a foreign nation. If something is in a neighboring county, that's their problem and there's not much we can do about it, even if it's going to have major impacts." He made this point to herald the accomplishments of Richmond County in forming a tight-knit coalition of neighboring counties to fight the LLRW site. With Wake and Chatham Counties, however, each hosted opposition movements at various times during the episode of contention, but they failed to sustain an effective working relationship. Moreover, although both the Chatham and Wake County commissioners officially opposed the LLRW site, they did not work together on the issue. At least two levels of geographic division presented problems for opposition organizers in the Wake-Chatham area. One was the division between the closest areas to the site on each county's side, and the other was the division between these areas and the rest of their respective counties.

The episode of contention in the Wake-Chatham area saw the formation of numerous groups opposed to the LLRW site scattered across the breadth of the two counties. However, these groups were short-lived and seemed to come in waves, with leadership and membership developing only to evaporate at a later point. I found only two activists who were active throughout the entire episode of contention. Other interviewees in the Wake-Chatham area confirmed this finding. One of the activists involved throughout the entire episode, Liz Cullington, said that organized opposition "was always sort of fading away and having to be reinvented. And it must have gone through about four incarnations. It had a succession of names."

Early Mobilization and Retraction

The first incarnation had formed more than a year before North Carolina even announced the four candidate sites. The group was called Chatham, Randolph, Moore Residents Against the Dump Site (CRM RADS) and formed in the town of Bennett, the self-proclaimed "friendliest town in North Carolina." This group raised more than $7,000 and had a membership list of over 1,000 people in 1989. It held rallies in the local school auditorium that attracted more than 100 people. CRM RADS mobilized in response to a list of 36 "potentially suitable areas" compiled by the Dames and Moore contracting firm in 1986 that was leaked to the *Raleigh News and Observer*. The group's founders believed that a site in the town of Bennett on the Chatham-Randolph border was a candidate site. CRM RADS members picketed outside the meeting of the North Carolina LLRW Management Authority, awaiting the announcement of the short list.

Surprisingly, none of the counties on the list of 36 became a candidate site. The closest site to CRM RAD's home in Bennett was more than 40 miles away on the Wake-Chatham County border. CRM RAD's leader, Jeralie Andrews, admitted that the group was relieved after the announcement of candidate sites because CRM RADS had "won its battle to keep the dump site out of the Bennett area." She explained that CRM RADS "wanted to push the entire dump site into Wake County because the nuclear plant is there" (Morris 1989). A resident of the area of Chatham County closest to the proposed LLRW site, which is known as Moncure, explained it this way: "Moncure has always been the redheaded stepchild of the county. They allowed anything to come in; we didn't have the political clout or a voice, whereas the rest of the county had a voice, and they wouldn't have allowed the things that happened in Moncure to happen in the rest of the county." When asked about CRM RADS, this Moncure resident said, "It wasn't in their backyard; they didn't have a dog in that fight."

To be fair, Andrews said that CRM RADS was not happy that part of the site's buffer area was in Chatham County, and she vowed to expand opposition to the siting process by helping organize the community near the candidate site (Morris 1989). CRM RADS organized a large rally and some organizational meetings in Moncure. Andrews addressed a crowd of 250 people and encouraged the Moncure residents, "Get yourself to Raleigh," and declared that "people have the power" (Denton 1989).

But rather than an expansion of opposition throughout Chatham and Wake Counties, there was what one activist called a handoff of leadership. From this activist's perspective, the CRM RADS leadership "just sort of handed it off to a new person and said, 'Well, you can be the president now,' ... and after that, [the group] sort of faded into nothing." Another activist said that CRM RADS "really just ceased to exist as a group; they didn't meet anymore." This organization could have provided a strong foundation for future mobilization against the LLRW facility. Yet while the Richmond opposition was circulating petitions and organizing FORRCE, the first organized opposition in the Wake County area was actually demobilizing.

A Second Mobilization Attempt Is Not Sustained

CRM RADS was succeeded by Central Carolina Residents Against the Dump Site (CC RADS) in January 1990. This group was a loose coalition of activists from Moncure, which was predominantly black, and white activists from other parts of Chatham County. CC RADS tried to mobilize Moncure residents through the churches, with some success. The group organized some rallies, candlelight vigils, and bake sales. One activist recalled, however, that "it was very tough getting the churches on board." Indeed, one member of CC RADS noted that her appeal to speak on the issue at several (but not all) churches was flatly, and sometimes rudely, denied by the pastors. A resident of Moncure said that "it seemed like the church didn't want to be involved, didn't want to associate with that." One CC RADS member, Liz Cullington, thought that the group suffered because "it never really was an organic organization. ... It wasn't a sort of homegrown movement," and "nobody would say, 'Now we're in charge.'" Kay Cameron, a CC RADS leader, said, "I don't remember a lot of leadership from the Moncure area. We tried to get it but couldn't. There was a minister from the Moncure area who was really active for a while, he had gotten a response, but I don't think it was a hang-in-there kind of group."

CC RADS also made an unsuccessful attempt to link the organization with opposition in Wake County. One man organized a group in Wake County called Wake Citizens Against Radiation (Wake CARE). Cameron wanted to join forces with Wake CARE, but the Wake County group declined. I could not find any evidence of the Wake group's public activities, and none of the activists in either Chatham or Wake County could recall anything about this organization.

Some of the original activists in CC RADS went on to oppose the site in an official capacity for a Chatham County committee charged with gathering technical information to prove the site unsuitable, the Site Designation Review Committee (SDRC). One CC RADS leader, Kay Cameron, followed this path. There was no formal ban against committee members pursuing citizen activism, she said, but "there was an understanding that this would not be prudent" to the committee's work. According to another CC RADS member, "The establishment of this committee absolutely depressed citizen activity, because two or three or more people who had been very active worked for this committee."

The SDRC became the primary force working against the LLRW site in Chatham County. All interviewees mentioned the influential role that this committee and its research coordinator, Mary MacDowell, played in opposing the site. However, as the SDRC took on a larger role in opposing the site, any citizen opposition that had been organized near the site dissipated. One committee member, John Graybeal, said, "I don't recall that there was any separately existing activist organization during the period of time that the committee was in existence." He thought that because of the SDRC, "some other folks who might have otherwise had a freestanding group weren't as active as they otherwise might have been." MacDowell said that the Chatham County SDRC held some of its meetings in Moncure "with hopes of getting those close to the facility to come," but "we almost never got citizens to come to the Moncure meetings. ... We found that residents closest to the site were very, very difficult to activate." Cameron recalled, "We wished that the people [of Moncure] would have been more active than they were. By that time, I was not going to those meetings. I didn't work with them closely."

Ground Zero: Yet Another Mobilization Attempt

Liz Cullington, one of the two activists involved throughout the entire episode of contention, recalled that at this point, "there was not much citizen effort, and after a few months we started over." CC RADS gave way to a new group of activists called Grass Roots Citizens Opposed to Nuclear Dumps, or Ground Zero, which began as yet another effort to draw the Moncure community into a larger opposition effort. The group held meetings at Liberty Chapel, a predominantly black church in Moncure. According to Cullington, this group formed when a professional organizer from Alabama moved into the area and "wanted us to start a group and get a grant and hire her as an organizer." Instead, the group chose two cochairmen: one from the Wake County side and one from Moncure. The professional organizer from Alabama "eventually just sort of gave up." The remaining group met once a month in Moncure at the chapel, held bake sales to raise money, and hosted a march from a church in Moncure to the proposed site, where a mock funeral was held.

But this activity quickly diminished as well. One of the founding chairmen of Ground Zero, Matt McConnell, noted that "the longer the process was drawn out, the harder it was to keep people involved. People seemed to be gung ho for a matter of months, but as it strung out, ... it became more of a commitment than people were willing to invest." He recalled, "It was hard, because we seemed to put a lot of effort in trying to combat it, but yet for some reason, we couldn't draw enough people in." Another group member described a meeting in Moncure where a woman stood up and said emphatically, "I've baked two cakes, I've marched, and I don't know what else I can do; it doesn't seem like I've made any difference." Cullington said, "I remember a lot of people being at the first meeting, but I don't remember a lot of people being there later on." Activist Nancy Tanguay believed that this effort "wasn't ever a successful match of banding together."

There were also racial barriers between the Moncure community and the landowners closest to the proposed site in Wake County. Cullington described the area surrounding the site as "divided socially, where people with different races were on two different tracks." Janet Zeller explained that "there was a county line division, and the black people were on the Moncure side and the white people were on the Apex side, and there might as well have been a wall from the ground all the way up to the sky in terms of people knowing one another or having any interaction at all."

The activists from Wake County were white middle-class professionals, while the Moncure community was predominantly low-income and black. One activist explained the situation: "Most of the leaders of this antidump thing were white, and we would try to arouse the folks in [Moncure]. And they were all black, and the leaders were all white and they didn't live right there. At one point, I had a conversation with them, and they said we can't get the church involved because they see this as the white man's politics, and we're having trouble staying involved because look at y'all—y'all are white."

Another Ground Zero member noted that there were organizational differences along racial lines as well. She explained that the black congregation at Liberty Chapel was used "to making decisions based on consensus and never really arguing," while all of us white people "were much more disputatious … we loved to get our teeth into things." After initial efforts to link residents in Moncure with those in Wake County, Ground Zero gradually focused its efforts exclusively on Wake County. One Ground Zero leader, Roseanne Edenhart-Pepe, who joined the group just months after these joint county efforts in Moncure, admitted that she was unaware of the group's original formation in Moncure. "I always thought that the Ground Zero that we were pushing forward was the original," she said. "But I found out five years afterward that there was a previous Ground Zero." Another Ground Zero leader, Pat Lehman, said that by the time she joined Ground Zero in this latter phase, "essentially it wasn't the same group of people" as it had been originally.

Ground Zero was able to coordinate more activities than previous groups had. Its members held fund-raisers, created radio ads, organized walkathons, had a float in the St. Patrick's Day parade, and spoke in front of some civic groups. Yet Ground Zero also found Wake County difficult to mobilize. It was never able to sustain an active membership, and it was never able to mobilize the entire county. A member of the Wake County SDRC, Don D'Ambrosi, said that "the vast majority of the population in Raleigh, which was the vast majority of the population in Wake County, really seemed to have very little interest in this." He reasoned that "this was something that was going to occur in a very remote southwestern corner of the county, and they didn't even know where it was. They'd never been down there. [To them,] it just wasn't there." A member of the Chatham County SDRC explained that "the big wheels that make the decisions all live in the luxury homes in North Raleigh away from the danger of it, and they don't have to live down there, so they didn't care." Edenhart-Pepe recalled similarly that people in the greater Raleigh area "just felt like this was a million miles away." According to Lehman, "It was hard to make people understand that this had any impact on them."

MacDowell said that "in Wake County, there was definitely just a tiny minority of people that were aware of this. . . . Ground Zero was just lost in the hugeness of the population there." She characterized the Ground Zero efforts in the following way: "They were really good people, but they didn't have good support. They were frustrated because they couldn't get a lot of people in the area active. These were all very intelligent people who were good at writing, talking, they did good interviews with the press. But a lot of the time, they were frustrated and just kept wondering, 'How are we going to make progress? We just can't get enough people out. What can we do when we are a relatively small group of activists?'" Ground Zero formed during what one activist described as the middle of the active opposition. But like the citizen opposition groups formed earlier, Ground Zero faded away before the siting process ended. One leader told me, "Frankly, . . . we burned out."

Just as none of these citizen efforts to link Chatham and Wake County opposition succeeded, the county governments also failed to adopt a common strategy. Each county government appointed a Site Designation Review Committee (SDRC), funded by the state. MacDowell recalled that the Chatham County Board of Commissioners charged their SDRC "to oppose the site," and she said the SDRC was "activist, proactive, fighting on all fronts." When she and other Chatham SDRC members approached the Wake County SDRC with the idea of uniting in opposition, MacDowell said the reception was "not friendly, not welcoming, not supportive." She maintained that "they didn't really want to listen to us." As MacDowell understood it, the members of the Wake County SDRC "were instructed by Malone [chair of the Wake County Board of Commissioners] to not be against the site. . . . They regarded our committee as very way out unacceptable, because our Board of Commissioners took a position against the site very early on, before I came on board."

Don D'Ambrosi, chair of the Wake County SDRC, confirmed this failed attempt at coordination: "Chatham folks got out in front in a negative way immediately. And that was bothersome to the Wake County Board of Commissioners." The Chatham SDRC "really wanted us to join them and fight the thing tooth and nail. And finally we told them as politely as we could that we have an obligation to the people of Wake County to look at this as objectively as possible ... we don't really want to cozy up here and do a joint committee to the point where everybody thinks philosophically we're as against this thing as you have stated that you are."

Consequently, even though both the Wake and Chatham SDRCs gathered information to show that the Wake site was unsuitable and lobbied the state government against the site, they never formed a working relationship that brought the two counties together.

Obstacles to Mobilization

Activists in the Wake-Chatham area were unable to overcome obstacles such as this split between the county governments, racial divisions among communities in the affected area, and detachment between communities near the proposed site and the wider Wake-Chatham population. Richmond, however, successfully overcame similar racial, geographic, and political obstacles. Yet the Wake-Chatham area faced some additional obstacles, frequently mentioned by area activists when explaining their mobilization challenge. For one, the proposed LLRW site was near a functioning nuclear power plant, a coal-fired generator, and several other polluting industries. Another factor was that a significant and growing proportion of the area population was highly professional and mobile.

The Carolina Power and Light (CP&L; now Progress Energy) Shearon-Harris nuclear power plant is located in the corner of Wake County that was selected to host the LLRW facility. Moncure, the area of Chatham County nearest the proposed LLRW facility, hosts a CP&L coal-fired generator as well as a Weyerhaeuser plant producing pressboard, a Honeywell plant manufacturing polyester, a brick factory, and a glue factory. LLRW activists in the area argued that this environment presented two different challenges to mobilization.

First, particularly on the Moncure side, most of the 600 residents nearest the proposed LLRW site worked for one of these already existing polluting industries. A Moncure resident explained, "Some residents worked at the jobs around there, and they didn't want to be political about it. Their employers would intimidate them. ... If they started getting a little radical, [their employers] would threaten their jobs in retaliation." MacDowell recalled that she was approached by citizens who said, "You can't challenge CP&L; CP&L is too powerful." She said that "people saw them as omnipotent in both Chatham and Wake Counties" and pointed to this as a major obstacle to mobilization efforts. "What we found was that residents closest to the sites were very, very difficult to activate. And one of the reasons was a great many of them worked for CP&L at the coal-fired plant in that same neighborhood or at the Harris Plant."

Edenhart-Pepe said that a lot of Wake County residents near the site were CP&L employees and "were either rabidly pro-nuclear . . . or they would say, 'We absolutely agree, but you can never tell anyone I said that, because I would get fired on the spot.'" She continued: "CP&L folks could never be active. They didn't want their name anywhere; they couldn't sign anything."

Second, longtime residents in the area had fought and lost many siting battles in the past. MacDowell spoke of a fatalism that hung over the community. She heard people say repeatedly, "They'll do it anyway—they'll put the nuclear plant there. They'll do it anyway; it doesn't matter, they're going to do it." She said the mobilization effort "was very discouraging, because we tried over and over again to let people know, and talk to people, and go door to door, and people just said, 'They're going to do it anyway; it doesn't matter.'" Priscilla Studholme, a member of CRM RADS, similarly noted about the response in Moncure that "a lot of people said, 'The government's going to do what it's going to do, and there's nothing you can do about it.' I mean, already in that part of the county, they've got horrendously polluting industry and there are health concerns . . . people were just beat down . . . they start believing that they can't have any effect."

Lehman said that native North Carolinians in the area were burned out. Residents would ask her, "We fought the siting of the reactor there and it didn't do any good, so what's the point?" The longtime landowners had weathered earlier siting battles concerning the construction of a man-made lake and a nuclear power plant. Opposition to each of these projects had failed, and each project claimed private property by eminent domain. One Ground Zero leader lived on a farm near several families that had held land in the area over several generations. She explained these landowners' reaction when she raised the LLRW issue: "I used to get on my horse to try to get them roused up. I asked them, 'Why don't you rise up against this?' And they would throw their hands up and say, 'They're going to do what they want to do; we can't stop them.' I'd say, 'Well, you can if you put some things together and do some things,' and they'd say, 'Yeah, we tried that with Jordan Lake and we tried that with the nuclear power plant, and it didn't work.' And that group began to slowly sell their land ... then that brought in developers, who started putting in these large subdivisions."

D'Ambrosi, an area developer, told me that these last remaining landowners were happy to sell their land at this point. He described the situation this way: "Now by and large, the old-line generational population ... couldn't have cared less. They're farm people; they had seen what had happened with the Research Triangle Park [when farmers had sold their land to developers for large sums of money]. ... So they were sitting out there with a 100- or 200-acre farm—a lot of these folks were sitting there saying, "Hmm, when is it going to be my turn to turn this thing into rooftops and make some money?"

Also, as environmental activist Lisa Finaldi noted, the area population "had already accepted a nuclear power plant," so the risk of an LLRW facility did not seem to be a serious threat. In fact, even some people who later became strong opponents of the LLRW site described their initial reaction this way. As John Graybeal of the Chatham SDRC put it, "I guess my initial reaction was that compared to the nuclear power plant that we have in our immediate vicinity here, a low-level radioactive waste site seemed like a kind of ho-hum situation." He said that only "over time, we came to realize that the siting for the low-level site created serious issues of potential water contamination."

While the Wake-Chatham area activists recognized the presence of a nuclear power plant and other polluting industries as an obstacle to mobilization, they perceived the transient nature of the growing professional population as their greatest challenge. According to the 1990 U.S. Census, between 1980 and 1990, the population of Chatham County increased by 16 percent, and that of Wake County by 41 percent. (Conversely, Richmond County's population decreased by 2 percent during this same period.) The population in Apex and Cary, the Wake County municipalities closest to the proposed LLRW site, increased 68 percent and 104 percent, respectively, during these 10 years (U.S. Census Bureau 2000). These townships were rapidly developing "bedroom communities" for the booming corporations of the Research Triangle. The Triangle J Council of Governments estimated that the Research Triangle region, which includes Chatham and Wake Counties, was adding 74 new residents each day (U.S. Census Bureau 2000). One statewide environmentalist explained that this rapid growth in the area meant that people churned through the community. "A lot of these people were professionals who had only been there a couple of years and then left in the middle of the night either for professional reasons or [because] they had enough money to move out of difficult situations." According to Mary MacDowell, "A lot of the population in Cary was transplanted from the North. So it wasn't like the fields and farms and rivers were their family homes. They lived in a suburb that was 10, 15 miles away from the nuclear plant and where the site was going to be. It just wasn't part of their radar screen of local concerns."

Environmentalist Lou Zeller recalled the effect the transitory nature of the community had on LLRW opposition groups in the area: "It's amazing: in Wake-Chatham, the people that we worked with to oppose the site in the beginning changed, the core constituency opposing the dump changed, so the people opposing the site in 1995 were a different group than those that had met on the issue just three years earlier." Pat Lehman of Ground Zero explained the effect this had on her community in Apex: "We didn't have any native North Carolinians in our group. Our community was tied to the Research Triangle Park and the corporations there. A lot of transients, people that are on a corporate climb … these people didn't plan to be there that long and had the capability to leave." Lehman's neighbor and fellow Ground Zero activist Roseanne Edenhart-Pepe provided a similar description: "The area we were living in was almost like tide coming in, tide going out. It was a transitional situation." MacDowell said, "The antinuclear activists in Wake County were just swallowed by the huge quantity of emigrants from the North and people who were just unaware and unconcerned."

Liz Cullington, who was active throughout the episode of contention, said this situation made it difficult to sustain active opposition, because "members of the public were always sort of floating through." Edenhart-Pepe explained that the opposition had "people at the beginning of the race, the middle of the race, and the end of the race." Nancy Tanguay, who became active with CC RADS and the original Ground Zero, recalled that "people for various reasons would leave a group and the character of the group would change."

One interviewee started listing people who got involved and then moved because their employers had transferred them to another part of the country. Matt McConnell of Ground Zero described this phenomenon as "waves of participation, because some of the people that I first met, after a year or so they kind of vanished." Pat Lehman used a similar analogy, saying that activists "washed in and then would move and take another job."

The three most prominent leaders of Ground Zero, McConnell, Lehman, and Edenhart-Pepe, moved in after the siting process began in 1989 and 1990 and all moved away before North Carolina halted the process. Lehman's husband was moved by his employer, and "we were there and then we were gone," she said. "And that happened with several other people. And that was the nature of the area." Edenhart-Pepe said she moved because she "got really tired of fighting every noxious LULU [locally undesirable land use] that came along, and it was part and parcel of living close to a nuclear site." McConnell explained, "Not knowing whether [the LLRW site] was going to be there or not . . . factored into our decision to move away from that area." He also said that the group's inability to sustain a high level of participation "burned a lot of people out." There were "just so few people, so much effort, that it just burned people out quickly."

Lack of Key Social Mechanisms

It is difficult to determine whether activists in the Wake-Chatham area could have successfully overcome such obstacles to mobilization. However, I could not find the same social mechanisms working in the Wake opposition that worked so well in Richmond County. The material above shows that the Wake-Chatham area did not experience an identity shift that linked previously detached social groups. Janet Zeller, who was familiar with the LLRW opposition in Richmond as well, said that the Wake-Chatham opposition never "joined in a common vision," and Lou Zeller said, "There was no community bond there."

It is not clear what social sites the Wake-Chatham activists could have appropriated to unify the various communities in the area, but some distinctions between Richmond and Wake-Chatham may be instructive. Janet Zeller recalled that "Richmond FORRCE knew how to party." She remembered "huge pig pickings" and "being fed to the max" in Richmond County. Lou talked of "a table eight feet long groaning with food, desserts, everything that you can imagine, fried chicken, barbeque." He said these community meals were "a real strength" in Richmond, because they "solidified community identity." Janet believed the meals "bonded people together." In contrast, Lou told me that in Wake-Chatham, "the groups never had any gatherings except meetings about the dump. They never got together in any social way—even for lunch." He said that "instead of mobilizing neighborhoods," the effort in Wake-Chatham "became more centered on the media campaign only and nothing in the community." Perhaps as a result of this, Lou reasoned, the Wake-Chatham area "was less active and less unified."

There is some evidence that activists in Wake-Chatham did not place a high value on attempts to bring disparate groups together. "From a participation point of view," Edenhart-Pepe said, "a lot of those meetings were just a big waste of time." She explained, "I think at times we erred in favor of trying to rally too many groups together and spent more time talking about it than actually doing it." Another activist complained of "meeting after meeting," claiming that some people thought "if we could all just get together on this, then we could get a lot more mileage, but the act of coordinating a large organization like that proved to be a distraction from the main goal of stopping the dump."

Also, Wake-Chatham activists frequently complained that the larger citizenry was distracted by other activities. In particular, multiple activists focused their complaints on basketball. One said, "Avoidance [in the community] was just rampant. People would tell me 'I'm so glad you're doing this, but I've got to go—the basketball game's on.'" Another lamented, "Our biggest problem was getting people to pay attention, and getting them to pay attention over a long period of time. They were just much more inclined to watch basketball. People weren't interested in anything except kids' school soccer or basketball, how Duke or UNC was doing." These complaints are striking, because Richmond County was fanatical about football, yet the activists in Richmond did not see this as a distraction or competition for community involvement—on the contrary, they appropriated the football team to rally supporters. Perhaps activists in Wake-Chatham could have appropriated basketball games to further their cause.

One final distinction between activists in Richmond and Wake-Chatham is the way they perceived their local political contexts. In Richmond, the activists suspected that their county commissioners had secretly volunteered for the LLRW site. They were even distrustful of the Richmond SDRC. Yet they still directed their energies toward their local officials. Leaders like Ariail and Quick conveyed a belief that any support from local officials, even if insincere, would help their cause. They perceived opportunities in the local political context.

This was not the case in Wake-Chatham. Even though the Wake Board of Commissioners officially opposed the proposed LLRW site, activists in the Wake-Chatham area never fully engaged these officials. When I asked Wake-Chatham activists about this, they expressed a belief that the county commissioners were not supportive, that they were beholden to CP&L, and that attempting to enlist their support would be a waste of time. Liz Cullington said, "On a governmental level, the Wake County commissioners never seemed to take much of an interest in this thing. . . . You know, I always assumed that it had something to do with the fact that they just had bigger things on their plate." Edenhart-Pepe maintained that "the county commissioners were just a useless bunch, . . . owned, bought, and sold by CP&L," and that attending county meetings "was a colossal waste of time." Tanguay told me she felt the county commissioners "weren't really listening to our concerns about the Moncure area, about the nuclear waste site." Lehman said that county SDRC meetings "were a joke," because officials were more concerned with politics than the issue of LLRW. She described the county SDRC meetings this way: "Politics was there. [My impression was that] those were people trying to go somewhere else politically. I don't even know who was there; I don't even know who the representatives were—that's how little impact they had. It felt like they were handpicked proponents. . . . They weren't active. It was about political climbing."

Richmond County activists also recognized that their county officials were not always motivated by pure motives, but they saw an opportunity to use political factors as leverage to their advantage on the LLRW issue. They invested energy in attempting to persuade even those officials they felt were hindering their efforts. Wake-Chatham activists, on the other hand, seemed to find local politics distasteful and therefore did not direct much of their energy toward county officials. They did not actively engage their local officials because they failed to perceive the opportunities in their local political context.

Mac Legerton, a statewide activist familiar with the situations in both Richmond and Wake-Chatham, believed that this was a significant failure. He argued that collaboration between local government and activists was essential to a successful mobilization: "I think collaboration is a good term. It is one who works with the enemy; one who consults the enemy. Good campaigns have to be collaborative. ... It is not so much a battle of good versus evil as a battle of how do we protect and promote our place and our people. And who's going to help us do that and who's going to hinder us, and how can we influence those we perceive as hindering that effort."

CONCLUDING REFLECTIONS ON OTHER CASES

How might the lessons of these two North Carolina cases apply to other candidate counties? The quantitative analysis in the appendix to this book found an unexpectedly negative relationship between the political leanings of elected officials on environmental issues and the number of collective acts of public opposition to the LLRW site proposal. Political leanings on environmental issues were measured as League of Conservation Voters (LCV) scores of U.S. House of Representative members serving candidate counties. The higher the score, the more supportive the representative's voting record on environmental issues. A closer look at cases in the upper and lower quartiles reveals no clear pattern on this measure. The LCV scores in the most active counties averaged 60 percent (with a range from 100 to 20 percent), and the scores in the least active counties averaged 61 percent (with a range from 100 to 10 percent). The LCV scoring system was intended to measure the presence of preexisting political opportunity structures favorable to an environmental struggle such as the LLRW siting.

The qualitative analysis of Wake and Richmond Counties illustrates that political opportunities are opportunities only if activists perceive them as such and act to create and cultivate such political advantages. Activists in Richmond County cultivated political advantages even in a conservative political context by fostering countywide support, directly pressuring their local officials, and defining the injustice as a property rights concern as well as an environmental concern.

Similarly, activists in New York's Allegany and Cortland Counties were able to mobilize active opposition in spite of an extremely conservative political context. The Republican candidates for the U.S. House of Representatives in these counties garnered 96 percent and 100 percent of the vote, respectively, in the election preceding the LLRW siting process. The activists in each of these counties successfully pressured their local- and state-level officials to oppose the LLRW site proposal and made use of a conservative property rights message. The following letter to the editor from a resident of Allegany County exemplifies a property rights theme similar to that identified in the Richmond County letters earlier in the chapter:

> In our country of freedoms, we've had a long history of governmental control of our property. ... Since the 19th century, our county and its landowners have been subject to the "eminent domain" of public works. From the Genesee Valley Canal to the railroads and trestles of the 1900s, to the gas lines of the 1950s, the Southern Tier Expressway of the '60s and '70s, to the high tension power lines of the '80s: in each case, some governmental agency had "right of way." I think it is important to remember that a government forcing its will on the people is not something new. This time, however, it's something deadly. (Szymanski 1989)

This letter from a Cortland resident also exemplifies the use of this property rights theme:

> I was brought up to love my country, but now as I grow older my feelings are fast changing. I am now learning to fear it and where they are leading us. Our government does not care what happens to us. ... I am a person, a wife, a mother, and soon to be grandmother. If one of you my fellow countrymen came on my property and refused to leave or tried to claim it for your own I could have you arrested. ... But our government can do this and I am powerless to prevent it. ... Somehow I find that rather ironic, that in our FREE COUNTRY we can be arrested for trying to protect our lives, homes and children. (Shevalier 1989)

Speaking to the conservative sensibilities of property rights on the LLRW issue is clearly one way to mobilize a community that is not otherwise environmentally engaged.

Just as activists can cultivate and even create political opportunities, they can also fail to take advantage of political opportunities in their midst. Wake

County had an LCV score of 90 percent and a county government that formally opposed the LLRW site proposal. Yet activists failed to fully take advantage of this seemingly advantageous political context. Of the seven counties in the lowest quartile of activism, five had an LCV score greater than 50 percent and a Republican vote in the 1990 U.S. House election of less than 50 percent.[1]

From Framing to Social Mechanisms and Process

The quantitative analysis in the appendix also found a significant positive relationship between the prevalence of an injustice frame in letters to the editor and active opposition. Indeed, Richmond County is joined in the upper quartile of activism by counties with the most frequent appearance of such frames during the siting process, and Wake County is joined in the lower quartile by counties with the least frequent expression of the injustice frame. I argue that this indicates the salience of the injustice frame in facility siting disputes. But more important, a frame is a social construction, an active process whereby activists work to define their cause during the episode of contention (McAdam et al. 2001). The significance of this variable indicates the importance of the mobilization process over preexisting factors such as political and civic structures.

The qualitative analysis identified social mechanisms that could be components of a mobilization process. In their introduction to the concept of the social mechanism, McAdam and colleagues defined *processes* as "regular sequences of such mechanisms that produce similar (generally more complex and contingent) transformations of those elements" (2001, 24). Does the combination of social appropriation, identity shift, and attribution of political opportunity constitute a regular sequence producing similar outcomes? Richmond County was joined in the upper quartile of active opposition by counties in Connecticut and New York. There is evidence from these cases that matches the Richmond experience.

---

[1] I consider seven counties to be in the lowest quartile of activism, even though there are only 21 cases total, because three of them registered eight collective acts of public opposition, the fifth-lowest tally. I consider these three plus the four counties that tallied fewer acts.

First, each of these cases appropriated key social sites. A review of the Cortland County, New York, opposition reveals appropriation of key countywide organizations such as the Chamber of Commerce, Christian Businessmen, Alliance of Churches, League of Women Voters, Rotary Club, Grange, union of university professors, and a garden club—all of which organized meetings on the dump issue, hosted speakers, and issued formal proclamations against the LLRW site proposal. Activists in Allegany County, New York, also appropriated countywide civic organizations, such as the Grange, Horticultural Society, Allegany Bird Club, teachers' organizations, school boards, churches, area universities, and volunteer fire departments. Activists in the counties of Hartford and Tolland, Connecticut, tapped into parent-teacher organizations and school boards, as well as the Chamber of Commerce in each county. However, the most important social site in the initial mobilizations of each of these counties seems to have been an association of real estate agents. For example, when the state publicly announced the candidate sites, a group of real estate agents in Tolland County immediately began to call all of their clients, established a telephone hotline, and organized a coalition against the LLRW site (Condon 1991b).

Second, activists cultivated an identify shift to overcome preexisting differences in the county and establish a unified county identity. The primary division Cortland and Allegany had to overcome was a split between people residing in the county seats and those living in the rural hinterlands. The original opposition to the LLRW site in Cortland actually formed in the rural area of the county. Leaders eventually moved the group's headquarters to a downtown office in the county seat to bring opponents from all backgrounds together. One leader of the opposition in Cortland, Paul Yaman, told me in a 2002 interview that "there was this coming together of a group of people who had never worked together before." Cortland County clergyman Delbert Wemple explained that at this point, all differences inside the county evaporated. "When it came to this issue, everybody was up in arms," he told me. "It didn't make any difference where you came from or what your persuasion was." In Allegany County, the best evidence of the countywide support for active opposition was this statement by a deputy sheriff who confronted a civil disobedience event against the LLRW facility and found farmers, doctors, college professors, teachers, and stay-at-home moms all linking arms: "I can't believe this. Did you see who was on the line out there? You've got everybody here" (Timberlake 1995, 29). The candidate counties in Connecticut, which were both bedroom communities for Hartford, had more significant racial and urban-rural divisions to bridge. They achieved such a bridge when an activist leader persuaded the Reverend Jesse Jackson to lead a march linking Hartford to outlying areas. In a speech, Jackson linked LLRW concerns to urban concerns in Hartford: "Only in moments like this we really see our oneness. We don't know it all the time, but we're really all one people. Just like Hartford is fighting its slums, Ellington is fighting a slum in the form of a nuclear dump. We should not be the tenants that turn the earth into a great slum" (Condon 1991a). Activists and governments in towns across Connecticut's Tolland and Hartford Counties joined in unique coalitions that shared funds and information and coordinated strategy.

Third, activists in New York and Connecticut worked to cultivate support in their local political context. Candidate counties in Connecticut developed the aforementioned coalitions of citizen groups and local governments to coordinate strategy. Within weeks of their formation, these coalitions successfully garnered active support from state and national representatives. Cortland County also formed a coalition linking citizen activists to local government officials. Citizen activists in Cortland successfully pressured the county government to hire activist leaders as full-time LLRW opposition coordinators. After opening this political opportunity, the Cortland opposition used its base in local government effectively to lobby state government. As we will see in the next chapter, Allegany County activists did not enjoy a friendly relationship with their county government. Although activists remained suspicious of local government throughout the siting process, they still worked to cultivate local and state government support for their cause.

The brief details of these cases in New York and Connecticut serve to complement the qualitative findings from the North Carolina analysis. The social mechanisms of social attribution, identity shift toward countywide unity, and attribution of political opportunity were actively engaged in counties that most successfully mounted active opposition to the LLRW site proposals. It appears that these mechanisms constitute a regular process that can facilitate the community mobilization of contentious politics.

Demographic Characteristics and Other Obstacles to Mobilization

The quantitative analysis of the facility siting professionals' model identified two unexpected demographic variables as significantly correlated with higher levels of mobilization: rural population and low educational attainment. The model had hypothesized that these factors would serve as obstacles to mobilization. Perhaps the deprived, rural, undereducated counties, such as Richmond, Cortland, and Allegany, targeted by the LLRW siting agencies, enjoyed a contextual advantage for mobilization. These rural communities provided a relatively easy environment in which activists could engage in the mobilization process outlined above. In these communities, social sites are easier to identify and more likely to be countywide, there is less diversity to obstruct the establishment of countywide unity, and local politicians are more accountable to constituents in a face-to-face manner. However, these advantages must be recognized as such by activist leaders. Five of the seven counties in the lowest quartile of activism were predominantly rural and met the profile outlined above. However, these demographic conditions are not a necessary condition for high levels of active opposition. Although Wake County seemed to suffer from a large, growing, diverse, and sprawling population, the Connecticut candidate sites were located in a similar demographic context and still managed to generate high levels of active opposition.

The Wake County analysis earlier in the chapter raised two additional preexisting conditions that deserve special attention. The opposition in Wake seemed to be hampered by geographic divisions and the presence of a nuclear power plant in the county. The successful mobilization of opposition in the Connecticut cases demonstrates that geographic divisions between urban centers and outlying areas can be overcome. The analysis of Richmond County and complementary evidence from the New York cases show that municipality and town-country divisions can also be successfully overcome. However, the characteristics of the counties in the lowest quartile of activism demonstrate that these divisions are extremely difficult to overcome. The only cases with a site that would directly and immediately impact two counties—Wake, North Carolina, and Hudspeth, Texas—are in the lowest quartile of activism. LLRW opponents in Hudspeth County generated the lowest possible score on my measure of active opposition, with just one collective act of public opposition to the LLRW facility in the first 100 days after the first act of opposition. The only other case with just one event on this measure was San Bernardino County, California, which shared Wake County's problem of a split between urban and sprawling outlying areas. Two other counties in the lowest quartile, Clark and Wayne, in Illinois, also struggled with irreconcilable differences between municipalities and outlying county populations.

Finally, the two remaining counties in the lowest quartile of activism share another characteristic with Wake County—they host nuclear power plants. Windham County, Vermont, and Nemaha County, Nebraska, generated very little active opposition to the LLRW facility proposal. Together with Wake, they are the only candidate counties hosting nuclear power plants. The Wake County analysis earlier in this chapter demonstrated that the presence of a nuclear power plant ensured that a significant number of people would enjoy the economic benefits of the utility industry, accept the risk of radiation, or be resigned to defeat based on the failure of past siting struggles.

This chapter demonstrates that an understanding of mobilized local opposition requires a fine-grained examination of the dynamic mechanisms and processes that facilitate mobilization. This finding supports an exploration of the active process of mobilization. Cases with high numbers of collective acts of public opposition to the LLRW sites demonstrated a concatenation of social attribution, identity shift to a countywide identity, and attribution of political opportunities. This sequence of social mechanisms was conspicuously absent from the least active cases. Geographic and demographic divisions, as well as the presence of nuclear power plants in the candidate county, may make such a sequence of social mechanisms more difficult to attain. The academic application of social mechanisms is simply an effort to make visible the way in which relationships among political actors interact with a political context and shape political outcomes. To understand the assertion of local political power in an implementation process such as the LLRW siting efforts, we need to understand not only what political resources are available in a community context, but also how human relationships shape the perception and application of that context. This chapter has detailed the ways in which such relationships shaped the *level* of active opposition across the candidate counties. The next chapter will explore the ways in which relationships shaped the *type* of active opposition across these cases.

CHAPTER 5

CRITICAL MASSES: DISRUPTIVE VERSUS CONVENTIONAL FORMS OF ACTIVE OPPOSITION

Collective acts of opposition come in many forms. In the spring of 1990, masked protesters on horseback in Allegany County, New York, engaged state troopers wielding billy clubs. The protesters were part of a successful and disruptive effort to prevent the state's LLRW siting officials from gaining access to proposed sites. Disruptive protest such as this, ranging from civil disobedience to violence, was relatively infrequent across LLRW candidate communities. Yet siting opponents in some candidate communities clearly made greater use of such tactics than others. What influences an opposition movement's strategic choice regarding the use of disruptive versus conventional forms of active opposition to siting decisions? The activity of local government in the siting dispute plays a significant role. In this chapter, I establish the relationship between collective acts of public opposition initiated by local government and depressed levels of disruptive protest. Carefully paired comparisons between candidate counties with different levels of disruptive protest reveal the way the nature of relationships between local governments and activist groups can either increase or moderate the level of disruptive protest. An understanding of social mechanisms such as brokerage, certification, and category formation reveals just how these relationships shape opposition activity.

PROTEST AS PART OF THE COLLECTIVE RESPONSE IN CANDIDATE COUNTIES

Scholars and observers of social movements in general, and opposition to waste facilities specifically, have tended to focus their attention on disruptive forms of activism. In the case of the LLRW siting implementation processes, such a focus would miss the vast universe of conventional collective acts of public opposition to the LLRW facility proposals and the important role local government played in the active opposition of most candidate counties.

Social movement scholars have focused on "disruptive direct action against elites, authorities, other groups, or cultural codes" (Tarrow 1998, 5). Conventional actions such as lobbying, litigation, negotiating with authorities, and participating in government institutions fall outside the definitional boundaries of the social movement. Della Porta and Diani argue that social movements are defined in part by "the frequent use of various forms of protest" (1999, 16). Andrian and Apter have joined disruptive acts to the very concept of "movement," preferring to use the term "protest movement" (1995, 1).

The literature on NIMBY, environmental justice, and antinuclear movements stays true to the boundaries of the social movement concept by focusing on the disruptive protest activity of grassroots groups. Szasz chronicles the "proliferation of local actions" opposing environmental harms in the 1980s, which he calls "disorder stories" (1994, 41). Prominent "disorder stories," such as protesters lying down in front of bulldozers to protest a hazardous waste landfill in Warren County, North Carolina, and civil disobedience events designed to thwart construction of the Shoreham and Seabrook nuclear power plants have figured prominently in the accounts of such movements (Bullard 1993; Eckstein 1997; McGurty 1995). The literature on local responses to LLRW facility proposals has also focused on disruptive collective acts by citizen activists. Kearney and Smith (1994) focus on citizens groups in Connecticut that organized patrols to intimidate state siting officials and blocked access to public meetings. Other authors have detailed mass civil disobedience events and even violence during the Nebraska siting process (Snowden 1997; Vari et al. 1994).

Yet analysis of collective action events gathered from local daily newspapers throughout the duration of the LLRW siting processes in 21 candidate counties reveals just 112 disruptive protest events over a cumulative total of 15,714 days of active siting processes.[1] This is one disruptive protest every 140 days, or 2.5 events a year.[2] This is hardly enough activity to justify the generally held belief that local opposition successfully thwarted each LLRW siting attempt (Albrecht 1999; English 1992; Gerrard 1994; Rabe 1994).

The social movement approach, with its focus on protest activity, is too restrictive to be instructive for LLRW siting implementation. The concept of "contentious politics" is much more appropriate for explaining these 21 cases. McAdam et al. define contentious politics as "collective political struggle" or, more specifically, "episodic, public, collective interaction among makers of claims and their objects when (a) at least one government is a claimant, an object of claims, or a party to the claims and (b) the claims would, if realized, affect the interests of at least one of the claimants" (2001, 5). This definition broadens the types of collective action events that qualify as part of the LLRW opposition. Contentious politics include the numerous letter-writing campaigns, petition drives, lawsuits, rallies, fund-raising events, informational meetings, lobbying efforts, and other conventional collective acts of public opposition that local communities used to fight proposed LLRW sites. This definition of contentious politics also allows for the possibility that a government, such as a municipality, can be a claimant.

Newspaper analysis of this broader universe of contentious politics reveals that disruptive protest is a small part of a much larger collective effort to oppose

---

[1] The duration of the siting process is measured from the moment a county is publicly named as a candidate site for a LLRW facility to the point at which the siting process is effectively ended. The cumulative total of days adds the duration of the siting process for each of the 21 candidate counties.

[2] Disruptive protest events are those in which more than two individuals express opposition to the LLRW facility by directly confronting authority figures or sites of authority with marches, pickets, civil disobedience, or other noninstitutional activities.

Total collective acts of public opposition	Cumulative duration in days of siting processes	Percentage of disruptive acts of protest	Percentage of local-government-initiated collective acts of opposition
914	15,714	12%	36%
		(N = 112)	(N = 331)

Table 5-1. Disruptive Acts of Protest and Local-Government-Initiated Collective Acts of Opposition as a Percentage of Total Collective Acts of Opposition in 21 Candidate Sites for an LLRW Facility

LLRW sites. Table 5.1 shows that over the same cumulative duration of days, these 21 cases exhibited 914 collective acts of public opposition, or one event every 17 days. Moreover, disruptive protest makes up only 12 percent of the total collective acts of public opposition. In contrast, more than one-third of all collective acts of opposition to the LLRW site proposals were conventional acts initiated by local government.

The relevant local government bodies registered official opposition to the LLRW siting proposal in all but 4 of the 21 cases. In many cases, local government clearly played a role in garnering collective opposition to the LLRW site proposal.

Variations in Disruptive Protest and Government Activity

Did the high frequency of local-government-initiated collective acts of public opposition depress the frequency of disruptive acts of protest? The social movement literature provides a solid theoretical basis to support this contention. Piven and Cloward famously argued that "political leaders, or elites allied with them, will try to quiet disturbances" by co-opting activists and channeling "the protestors into more legitimate and less disruptive forms of political behavior" (1979, 30). Tarrow summarized a wide swath of literature that demonstrated how disruptive forms of contention became "routinized into convention" through elite facilitation (1998, 104, 149).

It is important to note that local government participation in the opposition to LLRW sites was *not* co-optation. In most of these cases, the local governments were in fact claimants struggling alongside the local population against the state. The citizen activists and local government officials shared an identical goal: to prevent construction of an LLRW site in their county. There was no compromise as to the objectives of the collective action, but there may have been a compromise as to the method of collective action employed if local governments facilitated the forms of contention that the communities pursued.

Figure 5.1 considers the three counties with the highest frequency of disruptive protest in the first 100 days of the siting process (Allegeny, Wake, and Waldo) alongside the seven counties that had no disruptive protest in the first 100 days. The three counties exhibiting the most disruptive protest events in the first 100 days also had fewer government-initiated collective acts of opposition than any of the seven cases that exhibited no disruptive protest.

Table A.3 in the appendix presents event count analysis showing an inverse relationship between the frequency of government-initiated collective acts of public opposition and the frequency of disruptive protest events. Even though these results are statistically significant and explain some amount of variation, a small N and lack of variation in the dependent variable caution against concluding at this point. It is necessary instead to delve into individual cases to explain exactly how local governments affect the nature of collective opposition. Because different states pursued the siting process at different times and

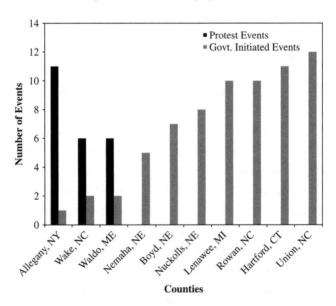

Figure 5-1. Disruptive Protest Events and Government-Initiated Opposition Events in the First 100 Days of the LLRW Siting Process Following the First Opposition Event

employed different implementation strategies, the best way to consider individual cases is to compare counties that vary in disruptive protest but shared the same siting process. Different states may also harbor variation in political culture that affects the propensity to engage in disruptive protest. Two of the counties with the most disruptive protest in the first 100 days, Allegany, New York, and Wake, North Carolina, are from states that also had less disruptive counties.

DISRUPTIVE VERSUS CONVENTIONAL FORMS OF CONTENTION IN NEW YORK

In 1988, holiday shoppers on Main Street in the small city of Cortland, New York, were greeted with signs that read "MERRY CHRISTMAS & HAPPY NU-CLEAR." Concerned citizens printed and distributed the signs in response to the December 20 revelation that Cortland County, along with a handful of other rural New York counties, was on a list of candidate areas to host an LLRW disposal facility. Nine months later, the New York State Low-Level Radioactive Waste Siting Commission compiled a "short list" of five potentially suitable sites. Three of these sites were in Allegany County, and the remaining two were in Cortland County. The residents in both of these counties expressed nearly universal opposition to the facility, and both communities mounted an intense and sustained campaign of collective action against the Siting Commission. During the 233 days that began when residents first learned their county was on the short list for an LLRW site and ended when the governor stopped on-site inspections, both counties generated 85 collective action events of opposition, or one event every 2.74 days. (These figures were gathered from event analysis of the local daily papers distributed in each county: the *Cortland Standard* and the *Olean Times Herald*.) This figure places these counties in a tie for the most frequent collective opposition generated among all U.S. counties facing LLRW siting proposals in the late 1980s and 1990s.

Yet the forms of contention activists employed in these two communities differed. The Allegany County opposition pursued a well-organized and disruptive campaign of civil disobedience that successfully kept state officials from surveying the land. The Cortland County opposition was less often disruptive and more focused on conventional forms of opposition than

Allegany. The two counties are remarkably similar along a wide range of variables thought to influence social movement activity, including demographic factors, political variables, measures of social capital, and issue frames.

The key explanatory difference between Allegany and Cortland Counties lies in the relationships that were cultivated during the siting process between the local governments and citizen activists in these two communities. Both counties were officially opposed to the LLRW site. However, the Cortland County officials worked with local activists, actually hiring several activists to coordinate opposition. This resulted in a unified strategy that embraced conventional and generally nondisruptive events. In Allegany County, local government officials were slow to act and unwilling to join in a common effort with local activists. This led to distrust between government officials and activists, despite their common opposition to the siting proposal. Ultimately, the local Allegany activists pursued a radical civil disobedience strategy that captured the support of the citizenry at large.

Different Forms of Contention

The character of the opposition in these two counties was very different. The high point of the opposition in Allegany County was a dramatic protest event that culminated in a confrontation between New York state troopers advancing on foot and masked protesters mounted on horseback. Both the protesters and the state troopers sustained injuries during the confrontation. The event successfully prevented the Siting Commission from gaining access to inspect the land and ultimately moved the governor to halt work on the sites. The high point of the Cortland County opposition came when actions they initiated and sustained led to a U.S. Supreme Court decision and a New York State law that eliminated the incentive and complicated the siting process of any similar facility in the future (*New York v. United States of America*, 1992, 505 U.S. 144).

These snapshots of successful opposition in each county represent the different forms of contention that activists in each county employed. Activists in Allegany County quickly adopted a disruptive strategy of intense civil disobedience confrontations with state officials. The Cortland County community was much more hesitant to employ such a strategy. Allegany had more than twice as many disruptive protest events as Cortland over the same

County	Percentage disruptive protest events (N)	Percentage conventional events (N)	Total collective opposition events
Allegany	32	68	85
	(27)	(58)	
Cortland	15	85	85
	(13)	(72)	

Table 5-2. Comparison of Cortland and Allegany Counties by Type of Collective Events Opposing LLRW Siting over the Same 233-Day Period
Note: Rounding error is present in all percentages in the table

time period, while Cortland exhibited a significantly larger percentage of conventional collective events than Allegany (see Table 5.2).

The arrest counts in the two counties present greater evidence of a difference in the character of opposition. The Allegany protest events resulted in 129 different individuals arrested, whereas in Cortland County, where protest events were less often civil disobedience efforts, police arrested just 29 persons. Perfectly reliable attendance figures for protest events are impossible to attain; however, when the local press coverage did estimate attendance at protest events, the average in Allegany County was 155 people, compared with 49 in Cortland. Cortland activists devoted most of their resources to disseminating information and hosting speakers at public meetings, creating technical reports that challenged the site selection, lobbying, and issuing legal challenges. During interviews, activists in both counties recognized what they considered the different tones and strategies of activism in the two counties.[3] The only other academic coverage of these two cases (O'Gorman 1997) also noted this difference.[4]

---

[3] For example, Richard Tupper, a county legislator who was very involved in the Cortland opposition, told me in a 2002 interview that Allegany County activists "really just weren't that involved" in the lobbying and legal challenges. Patrick Snyder, a Cortland lawyer, said that Allegany was "almost totally devoted to the civil unrest angle."

[4] O'Gorman used both of these cases as examples of local environmental movements that went beyond the NIMBY response to undesirable land uses to reflect a broader environmental concern. Although he presented the two counties as similar cases, he also noted what he called Allegany's "confronting" response and the Cortland "coordinating" response.

Insufficient Distinction among Common Explanatory Variables

The similarities between these two cases are as interesting as the differences. The "usual suspects" that social movement scholars use to explain different forms of contention all fall short here, because Allegany and Cortland Counties share nearly identical measures on (a) demographic variables, such as population, race, and income; (b) political variables, such as the party of locally elected representatives and the structure of local government; (c) social capital variables, such as the number and type of civic organizations, the protest experience of activists, and the role of key community groups in this struggle; and (d) a variable designed to identify the way the issue was framed in each county.

Repertoires of Contention: Demographic Variables and Activist Experience
The difference in the forms of contention employed by these two highly active communities fits squarely into the literature on repertoires of contention developed by Tilly (1978). He uses the term *repertoire* to identify a "set of routines" culturally limited by time and place that shape the contentious response of a group of people (Tilly 1995). The most obvious explanation for the difference between Allegany and Cortland is that these communities were not working from the same repertoire of contention. But both of these episodes of contention share identical lifespans. Activists in both communities were making choices as to a course of action within what some have called "the social movement society" (Meyer and Tarrow 1998). This was and is an era in which actors on all points of the ideological spectrum are familiar with and even comfortable using the full range of contentious forms, from petitions to protests.

However, time was only one of the determinants of repertoire constraint; place is just as significant. Yet it is hard to imagine two more demographically similar counties. The 1990 U.S. Census Data displayed in Table 5.3 shows similarities in race, income, and employment in farming and manufacturing.

Still, these demographics are just aggregates. Tilly noted the importance of knowledge, memory, and social connections in the population (1995, 27). What if a cadre of individuals in Allegany had significant previous experience with civil disobedience protest events? Interviews with the key people in the Allegany opposition showed that they were veterans of anti-Vietnam War demonstrations, the civil rights movement, and antinuclear protests. But interviews

Demographic statistic	Allegany County	Cortland County
Total population	50,470	48,963
Percent white population	98%	98%
Median household income	$24,164	$26,791
Percent of people 16 years or older employed in farming, forestry, or fisheries	5%	4%
Percent of people 16 years or older employed in manufacturing	21%	23%

Table 5-3. Selected 1990 U.S. Census Information for Allegany and Cortland Counties

with Cortland activists also revealed protest experience in events such as a civil rights march in Birmingham, the Freedom Summer Project, and antinuclear protests against the Seabrook nuclear power plant in New Hampshire.

Both communities also had small groups advocating violent confrontation. A mysterious group called Armed Citizens of Cortland County startled state officials with thinly veiled threats of violence at a large public hearing. Property destruction was conducted sporadically. This included crimping the propane gas line into the Siting Commission field office and dumping dead fish, skunks, cow manure, and fox urine inside the building (Nogas 1990; Nogas 1989). Allegany County residents were not above the use of roadkill either, as they dumped a dead skunk and cow manure into the ventilation shaft of the Siting Commission's mobile information trailer. A small group called the Allegany Hilltop Patrol planned but never carried out various acts of "ecotage" vandalism against heavy equipment working at the proposed sites (Lloyd 1995). Each community's repertoire also included more conventional (and peaceful) forms of contention. Cortland devoted millions of dollars and countless person-hours to legal challenges, lobbying trips to Albany, negotiations with state officials, letter-writing efforts, and petitions. A devoted group of academics and public officials in Allegany pushed for a contentious course that centered on proving to state officials that the county's sites were not suitable to host the waste. Though this effort ultimately took a backseat to disruptive acts in Allegany, it was nonetheless part of the repertoire. Tarrow wrote that contemporary movements drew on a repertoire of contention with "three basic types of collective action: violence, disruption, and convention" (1998, 104). Activists in both Allegany and Cortland

Counties were faced with this array. Allegany chose disruption, whereas Cortland chose convention.

Political Opportunity Structure: Governing System, Party in Power, and Policing

Many social movement scholars focus on political variables, often lumped under the heading "political opportunity structures," to explain episodes of contentious politics. McAdam neatly summarized the literature on political opportunities into four dimensions: (1) the relative openness or closure of the institutionalized political system; (2) the presence or absence of elite allies; (3) the stability or instability of elite alignments; and (4) the state's capacity and propensity for repression (McAdam 1996).

The openness of the political system refers to the governing structure. For example, Eisinger (1973) found that mayor-council systems, ward systems, and partisan elections offered residents access to the political system that helped moderate disruptive and violent forms of activism. The most comprehensive efforts to operationalize the next two facets of political opportunity structure used the party in power and the electoral margins of victory to measure the presence of elite allies and political stability (Meyer and Minkoff 1997). The final facet considers the importance of different policing strategies on citizen activism (Della Porta and Reiter 1998). Table 5.4 summarizes the first three facets of political opportunity structure for each county.

The two counties share identical institutional arrangements and electoral systems and had similar revenue numbers. During the siting struggle, Allegany and Cortland were clearly both safe Republican districts at the state and federal levels. Republican majorities governed both Allegany and Cortland county government. It does not matter whether we consider the Republican dominance in governance over these two counties a favorable or unfavorable aspect of the political opportunity structure—the counties are nearly identical. None of these factors evidences sufficient differences between the two cases.

Even if we enrich the concept of political opportunity structure to include elites that are not in elected positions, the two counties are still similar. The local daily papers covering both areas opposed the facility officially and publicly, as

Political variable	Allegany County	Cortland County
Form of county government	Council, no executive	Council, no executive
Electoral system	Partisan	Partisan
Council representation	By geographic area	By geographic area
County revenue	$32,378,000	$31,255,000
Party and percentage of vote won in 1988 by representative in the U.S. House	Republican 96%	Republican 100%
Number of registered voters	20,129	21,016
Plurality for president in each election from 1972 to 1988	All Republican	All Republican
Party and percentage of vote won in 1988 by representative in the New York State Assembly	Republican 100%	Republican 100%
Majority party in county government	Republican	Republican

Table 5-4. Comparative Political Variables for Allegany and Cortland Counties
Source: Information derived from New York State (1989–1990)

did the Chamber of Commerce, Grange, and several fraternal organizations in each community. In the press coverage that spanned the episode of contention, there was no discernible split or instability in elite opinion—the facility seemed to be universally opposed in both communities. Both county governments issued formal proclamations of opposition to the LLRW site.

The final dimension of political opportunity structure is the capacity and ability of the state to repress contentious activity. There is no doubt that the actions of the police who engage with activists affect contentious activity. Della Porta and Reiter (1998) have identified several relevant variables of policing protest, including the degree of force used, the timing of police intervention, the degree of police communication with demonstrators, police adaptability, and the degree of police preparation. Interviews with sheriff forces in both counties revealed an impressive degree of organization in dealing with protests and a considerable degree of sympathy for the community forces opposed to the low-level radioactive waste facility. Sheriff's deputies on both forces expressed solidarity with the citizens opposing the site. Both forces scheduled meetings with community activists inclined to protest and established "rules of engagement" before potential confrontations. Both forces expressed a desire to put the safety of the community above all else. In practice, this often meant letting protesters "vent" and successfully thwart state workers from conducting work on the prospective sites. Both forces showed physical restraint during confrontations and sought to avoid making arrests.[5] Activists in both counties commended the way their local sheriffs handled all confrontations.

Social Capital: Civic Organizations

Tarrow argued that "contention crystallizes into a social movement when it taps embedded social networks and connective structures" (1998, 23). Tarrow and other social movement scholars (McAdam 1999; Morris 1984) call such networks "mobilizing structures," but the concept is very similar to social capital. Civic organizations like clubs, churches, and fraternal groups are the essential elements used to evaluate the presence, absence, and amount of social connectedness in a community. Table 5.5 shows similar numbers of civic organizations and churches in each county. The civic organizations include fraternal lodges; sports and recreation clubs; public service groups; business, trade, and labor associations; veterans organizations; postsecondary education institutions; art clubs; and even bowling leagues.[6]

I use other measures to assess the mobilizing structures present in each county that would be more directly linked to the low-level radioactive waste struggle. Local daily newspaper accounts reveal that activists in both communities made contact with national antinuclear groups fairly early in the episode of contention and had contact with such groups with similar frequency. Both Allegany and Cortland had faced prior contentious environmental siting processes and formed citizens groups in response. Once again, this demonstrates that this element of the classical social movement

---

[5] The final protest was exceptional in this regard. At this event, New York state police used force to wrestle mounted protesters from their horses. Because this was the final protest event, it had no effect on future events in the episode of contention studied here.

[6] I used phone book listings to create an initial list of civic organizations (including churches) in each county. Next, I searched the Internet for local chapters of each organization listed in the back of Putnam (2000). Finally, I added any organizations that the local daily newspapers listed that had not emerged in the previous searches.

Indicator of social capital	Allegany County	Cortland County
Total number of civic organizations	87	100
Total number of churches and religious meeting places	36	37
Date of first contact with national antinuclear activists[a]	1/13/89	1/18/89
	Day 26	Day 31
Number of public contacts with outside antinuclear activists	13	14
Previous siting struggles took place in the county	Yes	Yes
Group established to oppose past siting efforts	Yes	Yes

Table 5-5. Comparative Indicators of Social Capital in Allegany and Cortland Counties
[a]This count is measured from the point at which the county first found out it was considered for a site. The count began eight months before the county residents found themselves on a short list of candidate sites.

agenda as it applies to the differing forms of contention is insufficient to explain the difference.

Issue Framing: The Way the Siting Struggle Was Perceived by the Public

Many environmental justice scholars explain the civil disobedience form of contention among low-income and minority populations over pollution issues by invoking an injustice or civil rights frame (Aronson 1997; McGurty 1995; Novotny 1995). I analyzed every letter to the editor concerning the LLRW facility in the local daily papers, then categorized each piece thematically by the predominant message the author was expressing. This compilation offers an understanding of the ways residents expressed themselves on the issue and of the messages they received on the issue. According to the work on environmental justice movements listed above, we might expect the dominant frame in Allegany to differ from that of Cortland, because Allegany acted predominantly with civil disobedience protests that were reminiscent of the civil rights movement.

In fact, the injustice frame appeared more frequently in the Cortland letters to the editor than it did in those of Allegany. Letters in Cortland expressed an injustice frame in 34 percent of the days during the siting process, whereas those in Allegany expressed an injustice frame in 20 percent. These two counties ranked first and second, respectively, among all the candidate counties in this study in their use of the injustice frame. The way the LLRW issue was framed, at least in the local newspapers, does not account for different forms of contention.

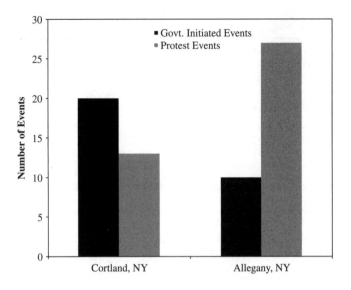

Figure 5-2. Total Number of Government-Initiated Opposition Events and Disruptive Protest Events during 233-Day Siting Process

A Process-Based Explanation: The Social Mechanisms of Brokerage, Certification, and Category Formation

These two cases essentially control for the variables of general collective opposition, available repertoire of contention, political opportunity, mobilizing structures, and strategic frames. They do, however, differ on the same variable identified in the first section of this chapter: government-initiated collective acts of public opposition (see Figure 5.2). Government involvement in collective opposition seems to depress disruptive protest.

If this is the case, how does government involvement in the opposition affect the level of disruptive protest? The answer to this question must come from a process-based explanation. More specifically, I aim to identify the key social mechanisms that link government involvement with the opposition, to less disruptive protest. As Hedström and Swedberg stated in their introduction to the use of social mechanisms, "mechanisms linking *explanans* and *explanandum* must be specified in order for an acceptable explanation to be at hand." This requires "deeper, more direct, and more fine-grained explanations" (1998, 8). I identify three key social mechanisms that work to explain the different forms of contention exhibited in Allegany and Cortland Counties. These mechanisms, which McAdam et al. (2001) have carefully defined, are brokerage, certification, and category formation.

Brokerage in Cortland County

Brokerage is "the linking of two or more previously unconnected social sites by a unit that mediates their relations with one another and/or with yet other sites" (McAdam et al. 2001, 26). In Cortland County, the mechanism of brokerage made unlikely allies of early rising hippie activists and conservative county officials. This new linkage led to a highly coordinated opposition effort of activists and local government.

During the week that Cortland residents learned their county was a candidate site for the low-level radioactive waste facility, concerned citizens, the town supervisors of several rural areas, and County Legislator Ted Law organized the first public meetings in the county (Conlon 1989a). Those gathered at these early meetings decided to call themselves the Coalition for Safe Communities (CSC), and they prioritized the following tasks: gathering information to show that the site would be a poor choice, establishing media relations, fund-raising, distributing information, and compiling petitions against the facility (Conlon 1989b).

During the second week, the citizen members of CSC broke away and started their own group, called Citizens Against Radioactive Dumping (CARD). A founding member of CSC and CARD, Denise Cote-Hopkins, told me in a 2002 interview, "Well, what the citizens did was immediately say, 'We're citizens, we don't trust any government whatsoever.' ... Those citizens just said, 'Nobody's to be trusted ... you're government, we're citizens.'" Several of these original CARD members are what other interviewees termed "the hippie element" in the opposition. These early activists had experience protesting against the Vietnam War, against the Seabrook nuclear power plant in New Hampshire, and with Greenpeace to save the whales. The early CARD membership possessed both protest experience and a distrust of government.

Two individuals served as brokers in Cortland to link the citizens, who were initially suspicious of the county government, and the conservative county legislators.[7] Denise Cote-Hopkins also told me that when Eleanor Ritter, a Republican town supervisor, found out about the prospect of a low-level radioactive waste dump near her township, she "felt that the first thing we needed to do was, we've got to make the county accountable." Ritter "didn't want this disconnect to exist between the county and the citizens." I also interviewed Ritter, who said she was disappointed with her first effort, as County Legislator James O'Mara dismissed her concern. However, her second effort found a powerful ally in County Legislator Ted Law. Ritter persuaded Law to come to the early meetings of CSC, and the citizens at these meetings urged him to get the county government involved in opposition. Within one week, Law ensured passage of a resolution opposing the LLRW facility (Conlon 1988). While CARD worked to mobilize attendance at the Siting Commission's first public hearing, the county government lent support at every turn—providing busing to the hearing, creating a task force on the issue, making information available, and directly confronting the commission with enumerated reasons why the county was an unsuitable site.

Ritter continued her contact with the citizens groups and successfully encouraged CARD to ask for the county legislature's support. CARD leaders asked that the county hire a full-time consultant to coordinate opposition. The day following the hearing, the county legislators voted unanimously to hire the most prominent original citizen activist in CSC and CARD, Cindy Monaco, to a full-time position heading the newly created and well-funded Cortland County Low-Level Radioactive Waste Office. Shortly after Monaco was hired, the county legislature also hired the environmental lawyer for the citizens group, Patrick Snyder, to work full-time opposing the facility.

Many activists and community members were surprised at the prominent role the county legislature took in opposing the dump and recognized that funding the opposition was an unprecedented government action. One resident I interviewed, Tom Milligan, who had a solid background in 1960s activism, had the following response to the hiring of full-time staff to oppose the facility: "That was a great surprise; our expectation was that they were just going to roll

---

[7] McAdam et al. define *brokerage* as "the linking of two or more currently unconnected social sites by a unit that mediates their relations with each other and/or with yet another site" (2001, 157).

over. They spent a lot of money, for them." I also interviewed County Legislator Dick Tupper, who explained just how remarkable these actions were for the Cortland County legislature: "It was pulling some teeth to get the county to put up $50,000 ... that's a lot in this county." Nevertheless, Tupper said that once the county government decided to support the opposition, "everybody was on board."

Brokerage in Allegany County

The early development of the Allegany opposition is very different. Instead of brokerage between citizen activists and local government, there was a seething rift. Although both the government and the activists opposed the LLRW site, they did not join their efforts. Brokerage did occur, however, between the mainstream citizens organization and a more radical organization devoted to civil disobedience.

In Allegany, the citizens group, which took the name Concerned Citizens of Allegany County (CCAC), organized without the participation of town or county officials. The Allegany County legislature released a statement announcing that it would take a position after the process of evaluating the sites unfolded (Dickenson 1988). The only government effort to organize opposition was undertaken by Allegany's New York state assemblyman, John Hasper, who called a closed meeting of about 30 local officials, university professors, and select citizens. When CCAC chairman Steve Myers asked to attend, he was refused entry. In a later interview (Hasper 1995), the assemblyman expressed the following concern: "The whole problem is that these groups form and people look and say are they serious or are they wackos. We don't need loose cannons." This fostered an already growing sense of distrust among citizens toward their elected officials. Steve Myers and his wife, Betsy, assumed that Hasper was trying to hijack the opposition and "make deals" behind the citizens' backs. Rich Kelley, another CCAC leader, was afraid that Hasper was trying to "take control and speak for the whole county" (Peterson 2002).

The gulf between the citizens opposition group and the county government could not have been any wider, and no one acted as a broker between the two. However, a different sort of brokerage occurred in Allegany that linked the CCAC to a more radical element in the community. Shortly after the county was named as a candidate site, Gary Lloyd helped organize a group called the Allegany Hilltop

Patrol. This group was a collection of avid outdoorsmen and backcountry woodsmen who met to commit to "protecting" the county with radical means, such as "vandalism for a good cause." By Lloyd's account, these were people "who were concerned about the dump, but weren't really interested in going the legal route with CCAC" (Lloyd 1995). At the same time, Stuart Campbell was getting more and more exasperated with the conventional forms of opposition he read about in the paper. As Campbell said, "It was clear to me that people had to do civil disobedience and that CCAC was doing all this other shit that wasn't going to stop it" (Campbell 1995b). As the membership of the CCAC became more disenchanted and distrustful of their local government, leader Steve Myers introduced Lloyd and Campbell to each other and gave them a forum to speak and recruit members for a civil disobedience approach at CCAC meetings. Myers's role as a broker was confirmed by Campbell, who said, "Steve Myers is absolutely key to having brought Gary and me together. I think that if either one of us had done that alone, probably nothing would have come of it" (Campbell 1995b).

Lloyd and Campbell formed a new group called the Allegany County Non-Violent Action Group (ACNAG). The group attracted people who were disenchanted with the conventional forms of contention that CCAC was pursuing. For example "Spike" Jones, a Vietnam War veteran, said, "I did not think my talents would help CCAC. There was nothing about what they were doing that even remotely appealed to me" (Jones 1995). Sally Campbell said, "I was never tempted to join CCAC, because that's just not my method of operating. I don't like to sit in meetings; I'm no good at fundraisers. It's just not my style of doing things" (Campbell 1995a). The original group also included a prominent CCAC leader, Jim Lucey, who would maintain contact between CCAC and ACNAG. At this point, as Stuart Campbell observed, "ACNAG was just the radical wing of CCAC." Later, CCAC would become "the legitimate wing of ACNAG"; they would be "two faces of the same thing" (Campbell 1995b).

Certification and Category Formation in Cortland County

Certification "refers to the validation of actors, their performances, and their claims by external authorities" (McAdam et al. 2001, 145). The coordination of CARD with the Cortland County government amounted to county certification of the citizen activists as legitimate actors with valid claims.

Category formation is the creation of "a set of sites that share a boundary distinguishing all of them from and relating all of them to at least one set of sites visibly excluded by the boundary" (McAdam et al. 2001, 143). The brokerage of CARD and the Cortland County government formed a unified category of opposition to the New York State LLRW Siting Commission.

County certification of CARD gave the group widespread legitimacy in the eyes of county residents. This transformed the group from a handful of citizens, identified in a 2002 interview by County Legislator Richard Tupper as "superactivists" and "sixties radicals who hate government," into a countywide organization with a large membership. The original CARD members had what activist Andy Mager called a "sort of in-your-face, 'fuck you' approach" in their engagement with the state and were just as apt to select a disruptive form of contention as they were to engage in conventional means. Yet county certification constrained CARD's activity. As McAdam and colleagues have noted, "Every polity also implicitly (and sometimes explicitly) broadcasts criteria for acceptable political organization, membership, identity, activity, and claim making," and "a certifying site always recognizes a radically limited range of identities, performances, and claims" (2001, 158). The county government recognized CARD's claims and actors, but it also constrained activity to conventional forms of contention. Meetings were soon altered from informal gatherings at members' houses to formal "Robert's Rules of Order" type meetings, with hundreds of residents gathered in public buildings. The group moved into a storefront office in downtown Cortland, purchased a new computer, and organized full-time staff. The original "hippie element" was soon outnumbered by retirees, businesspeople, and professionals. The new, larger membership adopted bylaws, elected new leadership, and consciously coordinated their action with the county.

When the county hired a CARD leader to serve as a full-time coordinator of opposition, the legislators did not choose a person from the "hippie element." Instead, they selected Cindy Monaco, a scientist with advanced degrees in both mathematics and environmental science. Dick Tupper, the Cortland County legislative chair, said that the legislature hired Monaco because her approach "was based on mathematics and science and it was not promotion. It was factual" (as quoted in O'Gorman 1997, 314). Monaco became the public voice of the opposition in Cortland County. Her scientific approach favored conventional forms of contentious politics, which rested well within the county government's certified parameters of action. Tupper described the county's preferred strategy to me in the following way: "The best response we could have was the governmental approach, which is the calm, well-financed, political process ... where we used Albany. We took advantage of Albany and Washington, and we used politics and our finances." Legislator Ted Law also favored the constrained and conventional opposition and thought Monaco fit right into this sort of opposition. He told me in a 2002 interview that she "wanted to fight" the facility, but "not disruptively." Monaco herself felt that "if you were viewed [in Albany] as one of the groups out there screaming and carrying on," the state would just "blow you off" (O'Gorman 1997, 322).

The social mechanism of category formation in Cortland brought CARD and the Cortland County government under the same umbrella of opposition to the New York State government. CARD's brokerage with and certification by the county government led to a tightly coordinated working relationship that joined these two social sites in a common "us" category that was defined against the "them" of New York State government.

In another 2002 interview, the publisher of the local paper, Kevin Howe, described Cindy Monaco as a kind of "bridge" or "connector" between citizens and local government. When she took the position with the county, she continued to attend the CARD meetings, and she attended all the sessions of the county legislature as well. County Legislator Dick Tupper also went to CARD meetings regularly. He told me that the county and CARD would correspond and coordinate activities sometimes on a daily basis. According to Duane and Betty Bonawitz, whom I also interviewed, the citizens group and the county-sponsored opposition became so close that CARD members volunteered at both the CARD office and the County Low-Level Radioactive Waste Office.

CARD cochair Paul Yaman and founding members Gary and Patti Michael each told me about the moment at which CARD decided to closely align itself with the county and follow Monaco's leadership. In the lead-up to a highly anticipated meeting with New York governor Mario Cuomo on the issue, a now small faction of CARD wanted to take a confrontational approach and stage a demonstration at the meeting. But in the end, Yaman recounted, CARD "finally agreed to ally ourselves with Cindy [Monaco] and, in turn, the local government at that meeting." The CARD representatives took "the county's approach ... which was [to] point out all these fallacies and really bring that into the light." The conventional approach sought to lobby Cuomo and negotiate a halt to the siting process. For the original CARD members, like the Michaels, "it was the activists themselves who got the meetings with Cuomo, but then Monaco, politicians, and that group took over." They saw Monaco as trying "very hard to influence group strategy" through "sympathetic leadership in CARD." From that point, Yaman said, the majority of CARD decided "all right, we're gonna follow the county line." The forces advocating convention won the intragroup competition over the form of contention. The opposition in Cortland would directly engage the state with lobbying and lawsuits supported by massive information gathering and education efforts.

Certification and Category Formation in Allegany County

Activists in Allegany County never benefited from the certification of their local government. The Allegany County legislature did ultimately adopt a resolution opposing the facility, but they did so five weeks after the citizens first learned their land was considered a candidate site (Dickenson 1989b). The CCAC leadership asked the county legislature repeatedly to commit money and personnel to the struggle. Fleurette Pelletier, action coordinator for CCAC, publicly requested that the county lend the support of its legal staff and budget to the opposition. She also requested that the county adopt legislation against the construction of nuclear waste facilities (Dickenson 1989d). The county attorney, James Sikaras, responded to Pelletier's formal requests weeks later by stating that it would be legally impossible to adopt such legislation according to the county charter. When it was suggested that the county amend its charter, Sikaras argued that this would be "complex and time-consuming" and would therefore be an ineffective way to counter the state. He also stated that the county was "not a vehicle for organizing opposition on any matter" (Dickenson 1989c). After Cortland County hired full-time employees to oppose the facility, Allegany residents asked their county legislature to do the same. Allegany County refused to follow Cortland's lead, and to the extent that they worked on the issue, the work went through the office of an administrative assistant with a modest budget of $5,000 (Dickenson 1989a). Whereas the Cortland County legislature hired a full-time environmental lawyer after the public hearing with the Siting Commission, the citizens group in Allegany lacked this kind of material support and public legitimacy.

The lack of local government brokerage and certification in Allegany affected category formation. The initial distrust and separation harbored by the citizen activists and their representatives seemed to be heightened during this early and critical phase. The only significant government-initiated activity explicitly excluded CCAC's leader. A CCAC cochair communicated the widespread belief that "the more prominent people were perhaps 'on the take,' perhaps had something to gain from this happening, that if they could bring this about without a big fuss, there would be political rewards" (Kelley 1995). The resulting category formation within the activist group excluded local government representatives. In fact, according to another CCAC cochair, "we figured that they [the politicians] were the enemy" (Myers 1995a). Allegany activists and local government in the county created mutually exclusive categories of opposition, leaving the activists to feel as though the entire struggle were up to them. This only reinforced a category formation that placed the citizens apart from their local government. The cochair of CCAC expressed this well: "The fight was up to us and that galvanized my position that if we were going to fight the battle we resolved not to let others co-opt us" (Myers 1995b).

McAdam et al. wrote that "decertification often emboldens insurgents to escalate their operations" (2001, 205). In Allegany, the distance between the CCAC and county officials emboldened the activists' resolve to confront the state and created an opening for advocates of direct and disruptive forms of contention. ACNAG founding member Tom Peterson explained, while conducting an interview (Beckhorn 1995), that there was "a political vacuum in the county," a lack of governmental leadership. The ACNAG founders wanted to provide leadership that would inspire communitywide action. The early ACNAG leaders, like Campbell, believed that the "people wanted to be led. They wanted a focus." Campbell knew that Allegany residents were upset about the siting proposal and disappointed in their elected representatives. Jim Lucey called this potential for widespread activism a "redneck ripple" of people primed to resist the state (Lucey 1995).

Notes from the first meetings show that ACNAG wanted to recruit 100 people that were willing to be arrested (Franklin 1989–1990). Spike Jones, a very charismatic former army recruiter, went to each local CCAC group to talk about civil disobedience. ACNAG leader Stuart Campbell obtained a map of the three candidate sites in Allegany County, which enabled him "to be like Paul Revere." Campbell went door to door informing people about the state's plans for the location of the LLRW site. ACNAG had 100 total members and 70 people willing to be arrested within a little over a month of the group's inception. Campbell said of this new and growing membership, "Those people were looking for things to do and ultimately all I had to say was 'Let's do this,' and if it made sense, they did it" (Campbell 1995b).

On May 31, 1989, the Allegany County sheriff arrested 48 ACNAG members when they linked arms around a car carrying members of the New York State LLRW Siting Commission. While this disruptive protest event was taking place, the Allegany County legislature was patiently convening in their chambers as they waited to meet with the Siting Commission. The legislators made absolutely no mention of the protest that was obviously escalating outside. They were probably the only people consciously avoiding the action. County office workers were leaning out of their windows, chanting, pumping their fists, and draping makeshift banners of support (Coch 1995). An overwhelming sense of empowerment swept through the crowd.

The support of the county office workers made a big impression on Stuart Campbell: "Workers in the courthouse actually hung out signs! Then I knew we'd scored. I knew that we had turned a corner." As he saw it, the people of Allegany County were "looking for ways to act," and ACNAG was "proving that we could act." From this moment, civil disobedience became the predominant form of contention employed by the Allegany opposition (Campbell 1995b). Gary Lloyd saw the effect of this early success on mobilization for even greater support: "An example was set; it was put in the paper; and all of a sudden, people, some of them anyway, thought, 'Holy Cow! Maybe we don't have to let this dump come in.' Then there was a geometric growth and it started spreading. The word spread and people said, 'Maybe we can do something'" (Lloyd 1995).

New members that committed to ACNAG and civil disobedience saw this first protest as a significant event. One woman said that at the first event, "I hadn't decided whether or not I wanted to be arrested at that point," but "after that I knew I was willing to be arrested. . . . Belmont changed me" (Zaccagni 1995). A senior citizen that joined ACNAG said that he was "influenced by the momentum that the group had gained" and his "admiration for the tactics, not only that they were nonviolent, but that they were highly creative and brave—and successful" (Warren 1996). Early CCAC leader Fleurette Pelletier said the membership in ACNAG was growing "because it was glitzy. There was a groundswell and people just went along with it. [It was] a fun thing" (Pelletier 1995).

Different Trajectories of Contention

When the LLRW siting process began in New York State, it would have been impossible to predict which of the two candidate counties, Allegany or Cortland, would engage in more disruptive protest events. The counties were nearly identical on the objective factors such as demographics, political opportunity structures, and mobilizing structures thought to be relevant to social movement activity. As the siting process developed, these two counties even showed similarities in the way the opposition framed the struggle against the LLRW site and the total number of collective acts of public opposition generated against the site proposal. The key difference lies in the process through which contentious politics developed in each county. Specifically, the social mechanisms of brokerage, certification, and category formation worked differently on the relationship between citizen activists and government officials in these two counties. This sent Cortland and Allegany Counties on different trajectories of contention.

In Cortland County, the mechanism of brokerage connected the county government to the newly formed citizens group opposing the LLRW facility. The two previously unconnected social sites coordinated their opposition. The county in effect certified the citizens' opposition, and the citizens formed a category of opposition that included the county government in the struggle against the external category of the state. Coordination between the county government and the citizen opposition put pressure on the county to devote resources to the opposition and limited citizen opposition to more conventional forms of contentious politics.

The Cortland opposition lobbied Governor Cuomo, the New York assembly and senate, and representatives in the U.S. Senate and House of Representatives. Lobbyists from both CARD and the county government successfully arranged an audit of the New York State Low-Level Radioactive Waste Siting Commission by the U.S. Congressional Budget Office; budget cuts for the Siting Commission; a state-sponsored National Academy of Sciences review of the siting process; a resolution opposing the siting process by the New York State Association of County Governments; and a new law governing the siting process that effectively halted the process. Cortland County made successful legal challenges to obtain technical information from the Siting Commission, force open meetings, mandate advance landowner notification of Siting Commission activity, and declare the federal law forcing states to construct low-level radioactive waste facilities unconstitutional. Throughout the course of the struggle, the county hired two more CARD leaders to coordinate opposition; an Albany lobbyist; Cornell University botanists, geologists, and hydrologists; and the premier environmental law firm in the United States. The county expenditures exceeded $1 million, and they successfully lobbied for reimbursement and ongoing funding in the New York State budget.

Some Cortland activists did attempt to organize more disruptive actions, such as civil disobedience. But this was a small group of activists that had lost the internal competition over which form of contention to employ. As the episode of contention progressed, this group got smaller and smaller, until a handful of radicals were getting arrested and rearrested at very small protest events. Less than 10 different people attended the last four protest events.[8]

In Allegany County, the citizen activists did not experience brokerage with the county government. Although the county government officially opposed the LLRW site proposal, it pursued this opposition apart from the newly formed citizens group. Thus the county failed to certify the burgeoning citizen activist opposition. This fostered distrust among the citizen activists, who formed a category of opposition that did not include local government. This distance between the activist community and the local government presented

---

[8] Interviews in 2002 with participants of Cortland protest events revealed a common sense of disillusionment. As one participant, Bruce Atkins, told me, "There were so few people there … it wasn't much of anything." Police video footage and participant interviews reveal that the largest protest events were disorganized and spontaneous. Another participant, Tom Mullins, expressed frustration with the event, stating that he was willing to be arrested "if it were truly, you know, a sit-down, but this is bullshit."

an opening for brokerage between the mainstream citizens movement and a more radical group advocating disruptive protest. Allegany ultimately pursued a disruptive trajectory of contention that was unconstrained by the conventional limitations of government-certified political activity.

Over the course of the siting process, ACNAG, the group advocating disruptive protest, expanded its membership and its activities and continued its impressive level of organization in subsequent protests. Over all the collective acts of opposition, 129 different people were arrested. At the final event, 500 people stood ready to be arrested. In each case, the Siting Commission stopped its attempts to gain access to the land.

DISRUPTIVE VERSUS CONVENTIONAL FORMS OF CONTENTION IN NORTH CAROLINA

Two counties besides Allegany, New York, exhibited more than five disruptive protest events in the first 100 days of the siting process: Waldo, Maine, and Wake, North Carolina. Each of these cases exhibited six disruptive protest events and just two government-initiated collective acts of public opposition in the first 100 days. Maine offers no less disruptive county to serve as a basis of comparison. Waldo is the only Maine county in this study. However, Wake was one of four North Carolina counties named on a short list of candidate sites. These three other North Carolina counties each displayed less disruptive protest than Wake and more government-initiated collective acts of opposition (see Figure 5.3). Qualitative analysis of these four cases shows the same social mechanisms that were at work in New York—brokerage, certification, and category formation—influenced the different forms of contention observed in North Carolina.

Rowan County, North Carolina

Newspaper accounts reveal that local government opposition in Rowan County, North Carolina, was immediate and dramatic. On the day the short list of four candidate sites was announced in North Carolina, local officials issued public statements communicating angry opposition. County Commissioner

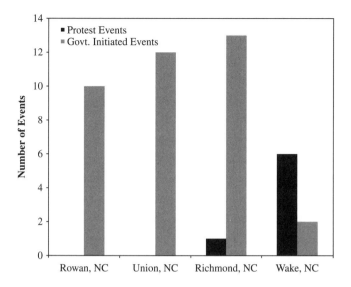

Figure 5-3. Disruptive Protests and Government-Initiated Collective Opposition in the First 100 Days of the Siting Process after the First Opposition Event in Each North Carolina County

Jamima DeMarcus declared, "I'm as mad about this as I've ever been in my life" (Bouser 1989a). She also vowed that "we will fight" the LLRW site "in every way we can" (Roberson 1989b). County Commission chair Newton Cohen was even more dramatic, predicting that the county government would be fighting the LLRW site "with every drop of blood in our bodies" (Roberson 1989b). Local government opposition in Rowan County was unequivocal, as Salisbury mayor John Wear communicated: "Our County says we don't want it, we don't want it under any circumstances" (Simmons 1989).

Local officials in Rowan County were also careful to distance themselves from the North Carolina state government and Chem-Nuclear Systems, the contractor charged with identifying an LLRW site. In this way, the county was forming a category of opposition that pitted "us," the county of Rowan, versus "them," the state and Chem-Nuclear. County Manager Tim Russell argued that state authority over waste disposal should extend only over state-owned lands. He declared that "if the state were responsible, they would put it on their own land" (Roberson 1989d). Commissioner DeMarcus made a clear distinction between state and local government, explaining that the siting process "is not a partnership with state government" (Roberson 1989b).

Cohen responded to Chem-Nuclear's projections of economic benefits and financial incentives for the host site by stating, "I believe they're trying to bribe us" (Roberson 1989d). After an initial informational meeting with Chem-Nuclear representatives, local officials refused to cosponsor a public forum with Chem-Nuclear because, as one official said, "It would inaccurately signal an allegiance with Chem-Nuclear" (Roberson 1989d). Local officials also refused to hold a joint press conference with Chem-Nuclear after this initial meeting.

While local officials distanced themselves from the state government and Chem-Nuclear, they actively reached out to citizen activists. After refusing to hold the joint press conference with Chem-Nuclear, DeMarcus began the process of brokerage with local environmental groups. The local newspaper reported that the commissioner had left the initial informational meeting to make phone calls to local citizen activists. DeMarcus said she was going "to marshal the forces" and make contact with several community activists. In the first week of the siting struggle, Rowan County officials invited leaders of the activist community to participate in a retreat along with the Chamber of Commerce to coordinate opposition to the LLRW site. After the retreat, environmental activist Ed Clement announced the formation of a coalition of environmentalists, county officials, legislators, and business leaders to "present a united front against possible storage of low-level radioactive waste in Rowan County." Clement communicated the feeling of mutual support among coalition members, remarking that "I think everybody's in a role of helping everybody else and assisting and cooperating and getting the job done." Importantly, Clement described the role of citizen activists as "support for county government's efforts." He explained, "I think government should be the primary player in this effort, supported by groups and citizens. Government is set up as an ongoing organization with management and with legal backing and monetary backing." The newspaper noted that he "praised Rowan County government's quick response on the issue" (Roberson 1989a).

The Rowan County government effectively performed brokerage with local environmental groups, certified local opposition to the LLRW site, and formed a unified category of opposition that grouped local government and activists together against the state and Chem-Nuclear. Under this arrangement, the county government took the lead in dictating the opposition activities. At the same time that local officials were framing their early opposition to the LLRW site in dramatic terms of a "battle" or "fight," they were warning against an overly "emotional" opposition strategy. The day after the state announced the candidate sites, Rowan County manager Russell advised that "facts, not emotion, will sway the [North Carolina LLRW] Authority and Chem-Nuclear" (Roberson 1989b). From the moment the Rowan site was publicly announced, local officials began to present technical reasons for disqualification including geology, groundwater flow, emergency response abilities, earthquake and hurricane history, and private property rights. Russell urged citizens interested in actively opposing the LLRW site to contact local environmental groups, which put volunteers to work identifying abandoned graveyards, Indian burial grounds, high groundwater tables, and other features that could disqualify the site. Citizens groups also coordinated conventional forms of opposition such as letter-writing campaigns, petition drives, and informational meetings. One week after the Rowan candidate site was announced, the Rowan County Board of Commissioners approved what they called a "blank check" to fund opposition. The commissioners charged the county manager and the county attorney to do whatever necessary to defeat the site proposal without a specified spending limit (Roberson 1989c). Rowan County used this "blank check" to hire Westinghouse to conduct an independent analysis of the site and an environmental law firm to pursue legal challenges to the site designation.

Union County, North Carolina

Opposition to the LLRW site proposal in Union County followed the same pattern as that in Rowan County. Government officials reacted with immediate opposition. Local newspaper accounts quote county politicians as declaring "this really upsets me" and "I was shocked and angry and dismayed." The press coverage summarized the local representatives' reactions as "promising to do everything possible to get [the LLRW site] out of Union County." John Munn, the county manager, promised that "we're going to make [the siting process] as tough as we can." Munn was also quick to engage in category formation, distancing the local government from the state siting process by declaring about the state, "If they want cooperation, they're not going to get it" (Voorhis 1989c).

In Union County, local government officials reached out to the activist community on the first day of the siting process. State Representative Clayton Loflin initiated the brokerage by advising Union County to approach an environmental group in neighboring Anson County for advice on "where to go from here" (Voorhis 1989c). Landon Scarborough, a leader in this environmental group, immediately notified Union County that his group would be "more than happy" to work with county officials (Voorhis 1989a).

Union County officials not only certified citizen activist opposition to the LLRW site, but also initiated a citizens opposition group. One week into the siting process, a group called Citizens Against Radiation Exposure (CARE) formed to oppose the LLRW site. Glenn Tippens, an officer of the newly formed organization, explained that local government officials encouraged him to form a grassroots organization. CARE included ordinary citizens and local elected representatives (Voorhis 1989b).

As in Rowan County, the Union County opposition often spoke in terms of battles and conflicts. The local newspaper warned that "county residents don't have to 'bear arms' yet to keep [the LLRW site] from coming, but you'd probably better keep your powder dry and your shot bag close to the door" (*Monroe (NC) Enquirer-Journal* 1989a). Despite this rhetoric, the form of contention proposed by the local government and followed by the citizen activists was conventional and aimed to use technical information to prove the site unsuitable. The local newspaper reported that Union County commissioners pledged to fight the LLRW proposal "technically" and made a plea for "better technical data." The county government agreed to pay whatever fees necessary to obtain this information, hired Westinghouse consultants to aid in this process, and asked citizens to help collect information and monitor Chem-Nuclear. When county officials announced this strategy at a public meeting, local activists "delivered a standing ovation" (Hoffman 1989a, 1989b).

Richmond County, North Carolina

Neither Rowan nor Union County engaged in a single disruptive protest event in the first 100 days of the siting struggle. Brokerage occurred quickly between local government and citizen activists in these communities, and mutual trust seemed readily evident. Activists in Richmond County were more suspicious of their local government officials, and organized citizen opposition formed independently of local government activity. However, the same three mechanisms of brokerage, certification, and category formation eventually worked to establish a nondisruptive collective response to the LLRW site proposal.

The initial public comments of local government officials in Richmond County reflected general opposition to the LLRW site proposal. State Senator Richard Conder "vowed to fight efforts to locate the dump in Richmond County" (*Richmond County Daily Journal* 1989b). The city of Hamlet, the municipality closest to the proposed site, officially opposed the LLRW site within one week following the site announcement. Councilman Bob Maloney declared that "Richmond County has no business getting into the business of handling this nation's nuclear waste" (Holland 1989b). J. Prentice Taylor, the chairman of the County Commission, assured citizens that "the Commission will do everything legally possible to protect the interests of the citizens" (*Richmond County Daily Journal* 1989b).

Yet the emerging citizen opposition to the LLRW site did not trust the local government to look out for their interests. One Richmond County activist I interviewed in 2003, who preferred to remain anonymous, described the public reaction regarding local government this way: "People were angry, I believe. I think everybody was angry and upset that this was all happening behind closed doors and that the public was the last one to find out about it and the deal had already been cut. The deal had already been made. People had already been paid off before we heard about it." Another described this prevailing belief in the following way: "This was orchestrated out of Raleigh, but with the consent of some of our elected politicians. This happens when a few think that they might be able to maybe gain something they want in return." Despite the public statements of local officials opposed to the LLRW site, most activists were convinced that Richmond County was a candidate site because local officials secretly welcomed the interest of Chem-Nuclear.

Concerned citizens formed an independent organization to oppose the LLRW site called For Richmond County Environment (FORRCE). The local paper reported that the concern emerging from the first FORRCE meeting was that local government officials "have remained silent on the issue." The new organization would "demand our elected officials to come forth in such an important time" (Holland 1989a). FORRCE members actively approached local officials with petitions and aggressively questioned them on the LLRW issue at public meetings. In yet another 2003 interview, FORRCE chairman David Ariail contended that actions like these solidified political support from local politicians. "The majority of elected officials seemed to follow [public opinion]. The populace was getting more vocal. It was becoming clear that, money or no money, the people didn't want it in Richmond County, and most politicians wanted to be reelected. ... Though the county commissioners were quiet at first, it was like everyone got new marching orders." He explained that "the suspicion that veiled our community from the beginning of an insider sellout was over. Now the county politicians were committing to spend whatever it took to legally fight the dump."

Ariail noted that as the local politicians came to openly oppose the LLRW, he became "the primary bridge between public opposition and local government." Ariail was a broker. When Richmond County held a public hearing on the issue, Ariail was offered a seat on the stage alongside elected officials. With acts like this, the local government was certifying FORRCE. When the county commissioners were forming legal strategy to fight the LLRW site proposal, Ariail was not only invited to the meeting, but also consulted for the "FORRCE position" on the issue. Richmond County's LLRW Site Designation Review Committee (SDRC) hired Peggy Coward, a FORRCE activist, to coordinate county opposition, because "she had ties to the community." One SDRC member told me in a 2003 telephone interview that Ariail garnered official certification because he "was able to pull a broad cross section of people together."

Coward noted that initially the local activists did not want to link up with the local government. She told me the activists didn't trust the politicians, "but I said, 'You've got to learn to use the system, you've got to play the system, no matter what. You've got to use what you got.'" Bobby Quick, a FORRCE leader, explained to me that he eventually "decided and convinced a lot of others that you're better off, instead of going after politicians and trying to get a vendetta against them or trying to get rid of them or anything like that, to just pull them over to your side. ... You're better off to get them working with you, or at least looking like they're working with you."

This effective brokerage between citizen activists and the county government, and the county certification of FORRCE, worked toward category formation joining FORRCE with the local government. FORRCE members came to serve on the county's SDRC, SDRC members and local government officials regularly attended FORRCE meetings, and the county sheriff provided deputy escorts for some FORRCE events. The wife of State Representative Don Dawkins became a fund-raising coordinator for FORRCE.

This unified effort ensured a more conventional form of contentious politics. County officials like the commission chair, J. Prentice Taylor, warned citizens to "avoid reacting to the situation emotionally" and advised the county to "stay calm and remain rational." The SDRC member told me that FORRCE leader David Ariail appealed to the county government because "he was rational and sound in his opposition ... not an in-your-face, radical, confrontational kind." Coward believed that she was chosen to work for the county in part because she was level-headed. She admitted that there were some in FORRCE "that wanted to make more noise and protest." However, Coward argued that FORRCE finally accepted the idea that "you can do more with sugar." She explained that civil disobedience would have bothered local officials and the general public "in a way that would have pulled them away from the sympathy that they had for us."

Wake County, North Carolina

The Wake County government, like the governments of Rowan, Union, and Richmond Counties, officially opposed the LLRW site proposal. However, unlike the other North Carolina counties, the local government in Wake did not inspire citizen activist support with its initial statements on the issue. The chair of the Wake County Board of Commissioners, Edmund Aycock, responded to the announcement of Wake as a candidate site by saying, "There will be considerable concern, but the fact is—there has got to be a low-level waste site" (Mather and Brooks 1989). This is a much more measured response than those of commissioners from other counties stating unqualified opposition and pledging to fight the proposal. Aycock said, "I don't anticipate that our county board will make an effort to block it, but I do think they'll make an effort to make sure that all criteria are adhered to and met" (Mather and Brooks 1989).

County officials and government appointees to the Wake County SDRC did recognize public opposition to the LLRW facility, yet they did not certify citizen activist opposition, nor did they group themselves in the same category as the activists. County officials interviewed said they did not attend any events organized by the citizen opposition. Unlike other counties, Wake did not appoint any members of the citizens opposition groups to the SDRC. When questioned about the lack of contact with the citizen activist community, another SDRC member I interviewed said the committee had decided, "We ain't going to have a circus." Another SDRC member made it clear that activists "weren't officially part of the group."

Like those in Richmond County, activists in Wake were suspicious of their local officials. However, in Wake, there were no identifiable brokers between the local government and the activist community. I interviewed several Wake County activists in 2003, several of whom asked to remain anonymous. One told me, "The Wake County Board was really bought off by the Carolina Power and Light Company and Duke Power and were very beholden to them." Another, Roseanne Edenhart-Pepe, also claimed that county officials "were owned, bought, and sold by Carolina Power and Light; that's the way Raleigh politics worked." Activists expressed frustration in their dealings with Wake County officials. One said, "The Wake County commissioners never seemed to take much of an interest in this thing." Another complained that the commissioners "didn't really want to listen to us. They were not sympathetic at all with our point of view." In other counties, the county-appointed Site Designation Review Committee served as a vehicle to aid brokerage between activists and county officials. In Wake County, the SDRC was seen by activists as either "a colossal waste of time," as Edenhart-Pepe put it, or an organization that gave residents a false sense of security and depressed citizen activism, from the viewpoint of activist leader Pat Lehman.

Without brokerage to ameliorate the distrust between the citizen activists and county officials, and with a building sense of frustration among activists regarding county opposition activity, activists were drawn to what one respondent called "highly visible" protest strategies. When asked what types of actions worked best in the LLRW opposition, activists pointed to marches, spontaneous protests at government meetings, and other disruptive actions that directly confronted government officials. As Lehman told me, "Whenever we were denied access to government, we tried to turn that into something highly visible, which worked great."

CONCLUSIONS

Active local opposition to LLRW facilities took many forms, from disruptive or even violent acts of vandalism and conflict to conventional political acts such as petitions and lawsuits. It was not only citizen activists who played a role in staging these acts, but also local officials. The relationship between activists and local government influenced the selection of strategic acts of opposition. Government-initiated collective opposition depressed disruptive protest when it effectively linked citizen activists and government officials in a common strategy of opposition. In less disruptive counties, the social mechanisms of brokerage, certification, and category formation sent activists and government officials together on a conventional trajectory of contentious politics. When brokerage was lacking between activists and local government officials, the government failed to certify the citizen opposition, and activists and officials were formed into different categories of opposition, as was the case in Allegany, New York, and Wake, North Carolina, where activists were moved to pursue more frequent acts of disruptive protest. The qualitative work on the 6 counties in New York and North Carolina described in this chapter lends credence to the correlation between high levels of government-initiated collective opposition and low levels of disruptive protest across 21 counties named as LLRW candidate sites. Disruptive protest was a rare occurrence in these counties because of the significant involvement of local government as a claimant in the contentious response to proposed LLRW sites. The full range of forms of active opposition played a role in obstructing implementation efforts of state siting authorities. In the next chapter, I consider the effects intergovernmental relations, both between local and state actors and among state actors, had on the implementation of the LLRWPA.

CHAPTER 6

RADIOACTIVE DECAY: IMPLEMENTATION FAILURE

The dramatic protests in Allegany County included more than just masked horsemen engaging state troopers. A group of elderly men and women under a banner reading, "GRANDPARENTS FOR THE FUTURE," braved the cold weather and chained themselves to a bridge—thus obstructing the Siting Commission. As state troopers began arresting this group, one of the women, Alex Landis, handed a folded flag to the county sheriff. It was the flag from the casket of her son, a World War II navy pilot. "If our freedom is taken away from me like this, I don't want this thing any more," she told the sheriff. "Help yourself" (Peterson 2002, 203).

Two days after the protest, Governor Mario Cuomo halted the LLRW siting process and admitted a change of heart on the issue, going from a siting advocate to a siting opponent. He even thanked opposition activists in New York for educating him on the issue. Cuomo insisted, "I don't want a dump anywhere" but instead "a more intelligent federal law" (Peterson 2002, 226).

He noted the "obvious and serious problems this federal mandate has created" and called on elected officials to recognize "the extraordinary interest of the involved communities" and to "reflect the desires of those communities" (Cuomo 1990). The governor eventually championed a successful lawsuit challenging the incentive structure of the LLRWPAA that was designed to force states to find new LLRW disposal sites. The flag that Landis had brought to the protest had made its way to Cuomo's desk, and he contacted Landis directly by phone: "This is Governor Cuomo. I'm sorry that you felt the government let you down. I'd like you to give us a chance to restore your faith in government. The siting commission came into Allegany County in an arbitrary way, with no consideration, no concern for the things the flag stands for. We're going to try to change that" (Peterson 2002, 231–232). Landis accepted the flag back, which state police returned to her in time for Memorial Day 1990.

Cuomo's transformation in response to local active opposition on the LLRW issue encapsulates two important intergovernmental dynamics affecting implementation: the influence of local interests over state-level prerogatives and the reverberations of this influence throughout the relationships among states in the U.S. federal system. In this case, Cuomo was responding to local political actors, and indeed, the very idea of a lawsuit challenging the constitutionality of the LLRWPA originated at the local level. The Supreme Court's decision on this issue and New York's refusal to implement the federal law (or at least stalling) dramatically altered the calculus of other states on the LLRW issue.

That local opposition to LLRW facilities thwarted implementation across the states is not a new finding. The stated purpose of the LLRWPA was to establish new disposal sites for LLRW. The fact that all such candidate sites were thwarted has led authors who have written about the LLRWPA to universally characterize it as an implementation failure attributable to "vigorous" (Vari et al. 1994, 64), "vehement" (Kemp 1992, 143), "determined" (Kearney and Smith 1994, 1), or "tremendous" (Rabe et al. 1994, 2) local opposition. However, few elaborate on just how the politics of local opposition bubbled back up through the federal system. A careful examination of the candidate counties reveals different paths to implementation failure. While it is true that none of the candidate counties in this study ever came to host an LLRW facility, they varied not only in the frequency, type, and timing of active opposition, but also in the amount of implementation progress made before siting processes were halted. The various paths to implementation failure across the local candidate sites interlaced through the federal system to create a systemic dynamic of failure among the states.

DEGREES OF IMPLEMENTATION PROGRESS

The U.S. Department of Energy (DOE) was charged by Congress under the LLRWPA to prepare annual reports detailing the progress of each state and regional compact along the following list of numbered milestones:

1. Create a siting procedure
2. Select candidate sites
3. Characterize candidate sites
4. Select site
5. Complete environmental assessment
6. Submit license request
7. Achieve license approval
8. Provide disposal

All 21 cases in this study met the first two milestones, but none achieved the final milestone. Thus the candidate counties vary from those that dropped out of the implementation process just after candidate site selection and before characterization to those that successfully achieved license approval for an LLRW disposal site before implementation halted. Figure 6.1 presents the variation between milestones 2 and 7 among LLRW candidate counties.

To what extent were the frequency, type, and timing of active local opposition related to progress along these milestones? The appendix to this book contains bivariate ordinary least squares (OLS) regression results considering the relationship between implementation milestone progress and each of the following: the number of collective acts of public opposition in the first 100 days after the first such act, the collective acts of public opposition during this time that were of a disruptive nature, the number of acts of opposition that were initiated by government, and the first day of local government opposition to the LLRW facility. Although the fact that there were

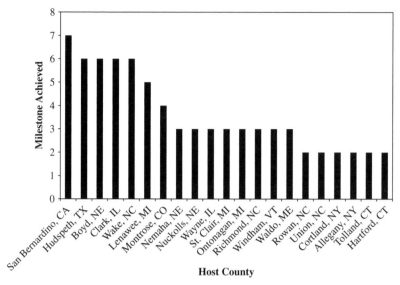

Figure 6-1. DOE Milestone Implementation Progress among LLRW Candidate Counties

only 21 candidate counties provides statistical problems for such analysis, the results do highlight the role of local activism in general, and local government in particular, in thwarting LLRW implementation and warrant a qualitative examination to reveal just how variations in frequency, type, and timing of active opposition affected implementation progress and the intergovernmental relationship between local and state actors.

It comes as no surprise that the frequency of active opposition has a significant negative relationship with implementation progress. The more actively opposed the candidate county, the more quickly the siting processes ended. Yet some scholars writing on LLRWPA implementation have paid particular attention to the type of active opposition—singling out disruptive acts as particularly important (Kearney and Smith 1994; Kemp 1992; Peterson 2002). And social movement scholars such as Gamson (1975) and Piven and Cloward (1979) have argued that movements employing disruptive protest are most likely to successfully achieve their policy goals. Yet the quantitative analysis does not find a significant relationship between the frequency of *disruptive* active opposition and implementation progress. This finding should lend credence to the call for an approach that casts a broader conceptual understanding of contentious politics—one that includes a place for more conventional acts of opposition such as those initiated by local governments. Interestingly, this analysis also reveals that the earlier and more involved a local government was in opposing the LLRW site, the less implementation progress the site made. The qualitative evidence that follows reveals many paths to implementation failure.

PATH #1: IMPLEMENTATION THWARTED BY EARLY AND OFTEN ACTIVE OPPOSITION

Table 6.1 displays the lowest quartile of progress along the implementation milestones. The six counties in the lowest quartile progressed no further than the announcement of candidate sites.

Although these counties represent the full range of types of active opposition they expressed—Allegany led the upper quartile of disruptive opposition, while Rowan, Hartford, and Union Counties were all at the bottom of the lowest quartile of disruptive opposition—they all demonstrated a relatively high frequency of active opposition during the first 100 days of their siting struggles. Four of these six counties (those in New York and Connecticut) were the most actively opposed candidate counties. The two remaining counties, Union and Rowan, North Carolina, were ranked seventh and eighth, respectively, in the number of collective acts of public opposition. How did local active opposition interact with the siting process in these cases?

| | | Lowest quartile | |
	Milestone	Collective acts of public opposition in the first 100 days following the first such act
Tolland, CT	2	51
Cortland, NY	2	44
Allegany, NY	2	37
Hartford, CT	2	35
Union, NC	2	24
Rowan, NC	2	21

Table 6-1. Activism among LLRW Candidate Counties in the Lowest Quartile of Milestone Implementation

There is evidence that in North Carolina, the site selection process was responding to the relative frequency of active opposition among the candidate sites. In a 2003 interview, State Representative George Miller, the sponsor of the LLRW compact bill and the champion of the LLRW siting process, admitted to me that "the politics of the situation removed the sites with opposition from further consideration." He said the siting process shifted its focus to "the most desirable, least active, but perhaps not the best site." An investigation by the Chatham County Board of Commissioners into the site-screening meetings of North Carolina's LLRW Siting Commission found that active opposition in candidate counties was discussed as much as hydrogeology, even though discussion of such nontechnical factors was supposed to be impermissible at these meetings. The investigation found that the amount of active opposition in candidate counties was considered "at every site screening meeting" (Farren 1992). The two counties with the most active opposition in the first 100 days, Union and Rowan, were dropped from the list of candidate counties before the implementation of milestone 3. Richmond County eventually was dropped from consideration after it met the third implementation milestone as a result of active opposition. As Representative Miller explained it, "The Richmond County site got removed, to the joy of the Richmond County people, for reasons other than whether it was a suitable site or not. It was really the opposition to the site that removed it from the list." This left Wake County as the preferred site in North Carolina because, as one consultant explained, "Wake County is simply not concerned except for a few people in the immediate site area" (Epley 1989, D009219). This sent Wake to progress to milestone 6 on a different path of implementation failure, described in more detail later in this chapter.

In New York and Connecticut, the entire LLRW siting enterprise halted before any candidate counties progressed further than the second implementation milestone. In New York, active local opposition generated by citizen opposition groups and local governments drew responses from state legislators and eventually the governor. Dick Tupper, the chair of the Cortland County legislature, believed that citizen activism effectively pressured state representatives. He told me in a 2002 interview that "the citizens group was a very integral part of the process, and I personally felt that we could take advantage of them keeping the issue alive, because that makes your local, Albany representatives very aware." Tupper pointed specifically to the response the local opposition drew from State Senator James Seward and State Assemblyman C. D. Rappleyea, both Republicans. Seward procured annual payments of $350,000 a year for four years to fund LLRW opposition in Allegany and Cortland. Rappleyea, the assembly minority leader and one of Albany's "power four" during the 1980s, helped gather Republican opposition to the LLRW siting process in both the New York state assembly and senate.

Tupper explained that Seward and Rappleyea garnered enough pressure in the legislative branch to influence Governor Cuomo's response to the LLRW issue. Tupper recalled, "Poor Governor Cuomo, whether he wanted to deal with the issue or not, he really had no choice; he was being pushed and he was being forced to deal with it." Cuomo himself, a Democrat, complained that he could get "no sympathy from the Republicans ... you couldn't have built [the LLRW facility] without the Republicans." He continued, "The Senate wasn't about to do anything that Rappleyea didn't want to do." Cuomo claimed that "there was no way out" of the controversy surrounding the proposed LLRW sites in Allegany and Cortland (Cuomo 1996, 8–10). Just two days after the explosive protest event in Allegany County involving protesters on horseback engaging New York state troopers, Governor Cuomo officially discontinued the siting process in Allegany and Cortland and began his constitutional challenge to the LLRWPA. It is impossible to disentangle the effects of the disruptive tactics employed by Allegany opponents and the more conventional tactics and local government actions employed by Cortland opponents. In retrospective interviews, Cuomo acknowledged the influence of both types of active opposition. He recognized that acts of civil disobedience in Allegany often compelled him to address the LLRW issue, and he acknowledged that he felt the political pressure of local government pushing up through state representatives (Peterson 2002). In the end, he instructed all state officials to respond to the local concerns variously expressed in the two counties with a cessation of the siting process (Cuomo 1990).

This same local-to-state ripple effect occurred in Connecticut. The day after the Connecticut Hazardous Waste Management Service announced the candidate counties, citizens in both counties formed opposition groups and organized a series of collective acts of public opposition. One week later, town and city governments were responding with formal statements of opposition to the proposed LLRW sites. These local officials attacked the state government for "imposing its will upon the towns" (Condon 1991b). State representatives for the affected areas responded within days, by joining the opposition and speaking at events organized by citizen opposition groups. Kearney and Smith, who compiled a detailed account of the opposition in Connecticut, argued that "with grassroots pressures building rapidly, local and state politicians were compelled to join in opposition" (1994, 6). State officials procured money to fund the opposition efforts of local governments. Ultimately, as in New York, implementation was effectively halted when the governor discontinued the siting process in the candidate counties.

Each of the cases in the lowest quartile of implementation progress exhibits the powerful impact of early and frequent active opposition. These cases in North Carolina, New York, and Connecticut each mounted relatively high numbers of collective acts of public opposition in the first months of the siting process. This activism forced a response from local and state elected officials. In North Carolina, the most active counties were dropped from consideration, while the least active county continued to meet implementation milestones. In New York and Connecticut, all of the candidate counties mounted frequent collective acts of public opposition. In these cases, the candidate counties won state funding for opposition efforts and eventually pressured the governor to halt the siting process.

PATH #2: WHEN ACTIVISM IS NOT ENOUGH

Active opposition exercised early and often was not always a recipe for an abrupt end to the siting process. The LLRW candidate counties that progressed to the highest implementation milestones in Illinois, Nebraska, and Michigan were also the most active candidate counties within their states. Figure 6.2 shows the implementation milestone progress and the number of collective acts of public opposition in the first 100 days following the first such act in each candidate county in these three states.

Among the counties in these three states, the highest number of collective acts of public opposition in the first 100 days was in Clark County, Illinois; Boyd County, Nebraska; and Lenawee County, Michigan. Yet each of these counties also progressed farthest in its state along the implementation milestones.

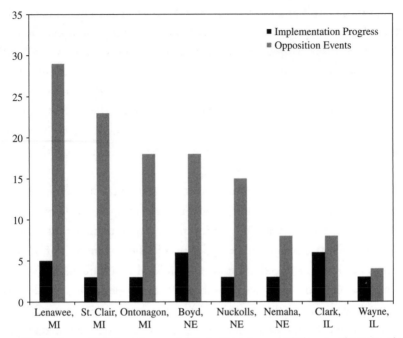

Figure 6-2. Activism and Milestone Progress in Illinois, Nebraska, and Michigan Candidate Counties

In both Nebraska and Illinois, the design of the siting process was responsible for this anomaly. Both states attempted to implement a "voluntary" site selection process. In Nebraska, Governor Kay Orr pursued implementation of an LLRW site for the Central Interstate Compact Commission under the condition that "the developer will not locate the facility in a community without the community's consent" (Thomas 1993, 49). However, this condition was never codified, and thus the definition of community consent was never established. Nevertheless, US Ecology, the contractor hired to site the LLRW facility in Nebraska, did follow a voluntary siting procedure. The contractor mailed out invitations to every local government in the state, asking if it would be interested in hosting an LLRW facility. US Ecology selected the three candidate sites in Nemaha, Nuckolls, and Boyd Counties from among those local governments that expressed interest. The announcement of the candidate sites generated significant active opposition in each of these counties.

In response, each county government formally opposed the LLRW site and asked to be taken off the list of candidate sites. The US Ecology invitation informed local governments that they could withdraw at any time. However, the invitation process did not distinguish between county governments and town or city governments. Nemaha and Nuckolls Counties were removed from the list on their requests. Boyd County was not, because the town board of Butte (which was within the borders of Boyd County) also responded to the invitation and continued to express a willingness to host the LLRW facility. Thus, even though Boyd County generated more collective acts of public opposition to the proposed LLRW facility than Nemaha and Nuckolls, and public opinion research showed that Boyd County residents overwhelmingly rejected the LLRW facility, it became the preferred site and progressed farther along the implementation milestones. Although the candidate site in Boyd County was not within the boundaries of Butte Township, and the other nearby townships of Naper and Spencer strongly opposed the site, US Ecology interpreted "community consent" as a single willing local government unit in proximity to the candidate site.

This interpretation and the selection of the preferred site in Boyd frustrated Boyd County activists and local officials who had already mounted significant opposition. One activist explained it this way: "When [US Ecology] come in here, everything was county. It went to the county board. Everything was county. It was Boyd County. Nuckolls County. Nemaha County. Everything was county. And that is what they meant by 'community consent.' County. But now, they don't have county community consent, so now they call it the Butte Site" (Thomas 1993, 178). Local active opposition in Boyd County did eventually generate its desired effect. As in Connecticut and New York, active opposition by citizen groups and local officials in Boyd County successfully pressured state officials to halt the siting process. Democratic gubernatorial candidate Ben Nelson took up the cause of LLRW opponents, going so far as to challenge incumbent governor Kay Orr to a debate on the topic in Boyd County. Nelson attacked Orr for failing to respond to the active opposition in Boyd: "We saw protests, arrests and canceled public meetings because of safety fears and Kay Orr was nowhere to be found … the state was represented in Spencer, but by the state patrol, not the governor" (AP 1990c). When Nelson defeated Orr and assumed the governorship, he called for a moratorium on the LLRW siting process in Nebraska (AP 1990b).

The Illinois Department of Nuclear Safety (IDNS) had a different approach to the identification of volunteer communities. It asked all counties in the state to issue formal notification if they did *not* wish to host an LLRW facility. Wayne and Clark Counties became candidate sites because they did not issue formal opposition. Bruce Rodman, a spokesman for the IDNS, said, "They are still contenders because of favorable environmental conditions and county boards which have not opposed the placement of a waste site in their area. If any of the counties register opposition later they will be dropped from consideration" (AP 1988). When the Wayne County board issued a statement of opposition, it was removed from consideration. However, when the Clark County board issued a similar statement, it remained a candidate site because the town of Martinsville, in this county, continued to express a willingness to host the site. Thus Clark County became the preferred site and continued to meet implementation milestones, even though it demonstrated more collective acts of public opposition than Wayne County. Two countywide referendums in Clark County showed that residents were against the LLRW facility by three to one. Although they could not remove their county from consideration at an early phase of implementation, activists in Clark County were instrumental in changing the final LLRW decisionmaking process.

The IDNS originally was responsible for both finding and approving an LLRW site, but activists in Clark County successfully pressured state lawmakers to take site approval away from the IDNS and put it in the hands of an independent commission. The commission ultimately judged the Clark County site technically unsuitable. Investigations by the Illinois senate later revealed that the IDNS had improperly given preference to the Clark County site in spite of known technical deficiencies.

The anomalous implementation results in Michigan were for different reasons than in Nebraska and Illinois. Each of the other cases considered thus far in this chapter demonstrates a process whereby local active opposition within the candidate county sent political pressure up to the state level, where implementation of the LLRWPA was ultimately halted. Governor James Blanchard was opposed to the LLRW siting process in Michigan, and the LLRWPA generally, even before the state named candidate counties. Blanchard had lobbied both Congress and the National Governors Association to revise the LLRWPA more than a year before Michigan named three LLRW candidate counties. Once the candidate counties were named, Blanchard announced that the state would file suit against the federal government to challenge the LLRWPA Amendments Act, arguing that it was "unconstitutional to require Michigan citizens to accept, against their will, the responsibility and liability— in perpetuity—for radioactive waste produced by private industry" (English 1992, 34).

Despite his public opposition to the LLRWPA, Blanchard went forward with the siting process in St. Clair, Ontonogan, and Lenawee Counties. However, he designed a process that made it impossible for a suitable site to be identified in Michigan.[1] James Cleary, commissioner of the Michigan LLRW Authority, said, "I doubt that any location in Michigan can meet the strict siting criteria for a low-level waste facility under State law. ... Our siting criteria go beyond the Federal requirements, and they may be prohibitive" (English 1992, 36).

Each of Michigan's candidate counties mounted significant active opposition to the LLRW site. Although Lenawee County mounted the largest number of collective acts of public opposition, the other two counties were dropped from the implementation process first. Unlike those in other cases in this study, citizen activists and local leaders in Lenawee County did not seem to voice much concern regarding their selection as the preferred site. Instead, they realized that the siting regulations in Michigan would soon halt the siting process in their county as well. When Michigan announced that Lenawee was the sole remaining candidate county, a town supervisor for the municipality closest to the proposed site expressed confidence that the site would be eliminated because it violated technical criteria. The state representative for the county declared that the site "will be eliminated. I am confident of that. It's just a matter of time" (AP 1990d). An activist leader said, "Truly, I don't think there is a site in Michigan" (Frownfelder 1990).

Even Cleary, who was orchestrating the site selection, hinted that Lenawee would soon be off the list. Whereas siting agencies in other states refused to publicly admit technical weaknesses in any candidate sites, Cleary outlined the precise aspects of the Lenawee landscape he thought would disqualify it as a candidate site. Three months after Lenawee was named the preferred site, it

---

[1] For more on the role of the governor in stalling the Michigan siting process. see Hill and Weissert (1995).

too was dropped from consideration. Governor Blanchard halted the siting process and argued that the Midwest Compact Commission had to find another host state, because Michigan did not contain any suitable locations for LLRW disposal. It seems that citizens, local officials, and the siting authorities all knew that the siting process in Michigan was not intended to succeed. The advancement of Lenawee County along the implementation milestones was meaningless, because all parties involved knew it would never be licensed. After Michigan dropped the Lenawee site, Governor Blanchard declared, "Our goal is to keep a low-level radioactive waste dump out of Michigan" (Mostaghel 1994, 402).

PATH #3: IMPLEMENTATION FAILURE IN THE ABSENCE OF ACTIVE OPPOSITION: THE VULNERABILITIES OF DEVOLUTION

Actions like those of Governor Blanchard in Michigan or Governor Cuomo in New York do not simply affect in-state implementation of the LLRWPA; they reverberate among all the other states as well. As local opposition inspired the opposition of leaders in state government, siting processes even in candidate counties with relatively little active opposition began to falter. Table 6.2 displays the upper quartile of candidate counties according to implementation milestone achieved. These five counties moved closest to actually hosting an LLRW disposal facility. San Bernardino County, California, actually received a license for an LLRW site. The other four counties submitted license applications before the siting process was terminated.

	Upper quartile	
	Milestone	Collective acts of public opposition in the first 100 days following the first such act
San Bernardino, CA	7	1
Boyd, NE	6	18
Wake, NC	6	8
Clark, IL	6	8
Hudspeth, TX	6	1

Table 6-2. Activism among LLRW Candidate Counties in the Upper Quartile of Milestone Implementation

With the exception of Boyd County, Nebraska, which was discussed above, these counties were all in the lowest quartile of active opposition. Generally, the least actively opposed counties progressed farthest toward successful implementation. Yet none of these candidate sites came to host an LLRW facility. Why, even with the absence of significant amounts of local active opposition, did the implementation of the LLRWPA ultimately fail?

Analysis of Hudspeth, Wake, and San Bernardino Counties shows that in the final stages of implementation, active local opposition was not necessary to derail progress. In each of these cases, state government actors intervened to thwart implementation in response to a new political calculus of LLRW disposal. As other siting authorities across the country halted implementation, the perception emerged that the remaining candidate sites would become national rather than regional disposal sites. The policy trajectory of high-level radioactive waste disposal quite possibly reinforced a fear among states of hosting a national site. In 1987, Congress abandoned a technical search for regional high-level radioactive disposal sites with passage of the Nuclear Waste Policy Amendments Act. Known to state residents as the "screw Nevada plan," this law singled out the Yucca Mountain site in Nevada as the only candidate site for the nation's high-level waste.

State government actors reacted to this new perception by first slowing, then ultimately stopping, LLRW siting processes across the country. The siting process in Texas was among the first to reflect the emergent fear among states of hosting a "national" LLRW facility. In 1986, Hudspeth County, Texas, was the first county named as a candidate site for an LLRW facility by a state siting process. Over the entire duration of the siting process between November 1986 and January 1991, Hudspeth County generated just 29 collective acts of public opposition. It was two elected judges, the governor, and the Texas legislature that ultimately thwarted the LLRW siting process. Luther Jones, an elected judge on the El Paso County Commissioners Court,[2] levied a special sales tax on telecommunications services in El Paso County to fund a series of lawsuits against the Texas LLRW Authority and hire lobbyists. The proposed LLRW site was located 16 miles from the El Paso County line. He was acting on his

---

[2] Judge Jones was an elected constitutional county judge, which is the office vested with executive responsibilities in Texas county government.

belief that the Hudspeth County site "would become a national dumping ground" (Scanlon 1987). Ultimately, Judge Jones found an ally in Judge Bill Moody of the 34th Texas State District Court, who ruled that the Hudspeth County site was unsuitable and issued a permanent injunction halting the siting process. Judge Moody likened the Texas LLRW Authority's siting process to Iraq's invasion of Kuwait and the opposition efforts in El Paso County to the coalition of nations fighting against Iraq (*Nuclear News 1991*). Before the Texas LLRW Authority could appeal this case, newly elected governor Ann Richards personally urged the authority to abandon the Hudspeth County site. She supported Judge Moody's contention that the site would present unacceptable risks to public health and the environment (DOE 1992). Shortly after this statement, the state legislature amended the Texas LLRW law to remove the site from consideration.

By 1992, Governor Richards in Texas was joining a growing wave of state governors stalling or even thwarting implementation of the LLRWPA. Governor Cuomo and the state legislature in New York had halted the operations and reduced the funding of the state siting authority. Michigan governor James Blanchard had declared his opposition to the LLRWPA, arguing that congressional action was needed to reduce the number of LLRW facilities currently planned across the country. A lack of siting progress in Michigan led to its expulsion from the Midwest Compact. Nebraska governor Ben Nelson had filed suit against the Central Interstate Compact Commission to remove the Boyd site from consideration. Local citizen activists from the opposition movements in Allegany and Cortland, New York, had traveled to both Nebraska and Connecticut to support burgeoning local opposition there. In Connecticut, the governor and general assembly had terminated siting activities. The Illinois legislature had disbanded its siting commission.

Other state siting authorities may not yet have taken as boldly defiant actions as these, but they were stalling implementation. Pennsylvania, host state for the Appalachian Compact, for example, never reached the site selection phase of implementation, because the siting process worked by applying disqualifying criteria in multiple layers. Each application of disqualifying criteria removed a larger percentage of the state from consideration, and the siting process never produced candidate sites. Massachusetts also stalled implementation by applying disqualifying criteria. New Jersey and Ohio each stalled implementation by delaying creation of a siting process with numerous exercises by boards and commissions to research successful implementation. The 1992 DOE report on the siting process stated that "the majority of States' progress has slowed due to political and legal challenges and public opposition," and that state action had been "insufficient to culminate in the ultimate goal—providing disposal capacity by January 1, 1993" (2, viii).

State inaction under the strict milestones outlined in the 1985 amendments to the LLRWPA was not without its consequences. The states hosting the existing LLRW facilities—Nevada, Washington, and South Carolina—could levy surcharges on or even deny access to waste from states that failed to demonstrate implementation progress. By 1992, these "sited states" had begun to levy surcharges and exclude states perceived as failing to make adequate siting progress. The 1985 amendments also stipulated that any state that could not provide for disposal by January 1, 1996, would have to "take title to the waste, be obligated to take possession of the waste, and shall be liable for all damages directly or indirectly incurred by such generator or owner as a consequence of the failure of the state to take possession of the waste." In 1992, as it looked increasingly unlikely that any new facility would be completed by 1996, state leaders anxiously awaited the outcome of a U.S. Supreme Court decision on the constitutionality of the LLRWPA.

The idea of a constitutional challenge had its origins in the local opposition movements of the New York candidate sites. Although Governor Cuomo brought the suit and declared that he would set out "to kill the federal statute," he was initially skeptical of the likelihood of success of a constitutional challenge, pointing to the fact that the federal courts tended to side with the federal government on challenges stemming from the 10th Amendment reserved powers of the states (Peterson 2002, 226). The case was dismissed by a U.S. district court—a decision that was reaffirmed by the Second Circuit Court of Appeals. To the surprise of Cuomo and many others, the U.S. Supreme Court granted New York's petition to hear the case. Connecticut, Michigan, and Ohio filed amicus briefs in support of New York. *New York v. United States* marked a conservative turn on the Supreme Court with a new interest in the 10th Amendment—a resurgence of states' rights and the placement of limits on federal power. In this case, New York argued that the incentives enacted under the 1985 amendments to the LLRWPA, including surcharges, the exclusion of waste disposal by states with existing sites, and the "take title" provision, infringed on its reserved powers under the 10th Amendment, complaining that "the states are simply ordered by Congress to take part in this activity." During the hearing, Justice Sandra Day O'Connor sarcastically called the take title provision "a pretty clever scheme" and suggested that "maybe Congress can require the states to each take over a share of the national debt" (Greenhouse 1992). In a six-to-three decision, the court found that surcharges were constitutional under the tax-and-spending powers granted to the federal government, the exclusion of waste was constitutional under congressional authority to regulate interstate commerce, but the take title provision was an unconstitutional attempt by the federal government to "commandeer" the state governments to participate in a federal regulatory program. In the end, the court left the LLRWPA and its amendments in place but struck down the take title provision. The decision was issued in June 1992.

While *New York v. United States* removed the sharpest teeth from the LLRWPA, falling waste generation levels also worked against implementation progress. When Michigan governor James Blanchard was taking steps to halt the siting process in his state and depart from the Midwest Compact, he argued that the LLRWPA was flawed because it forced the creation of far more LLRW facilities than waste volumes and the economics of disposal necessitated. Both he and Cuomo argued that the LLRW disposal problem would be better solved by just one site—possibly serving the disposal needs of all types of radioactive waste. A 1992 GAO report presented both evidence of the falling volumes of LLRW and the emerging view that fewer, more centralized disposal sites would be adequate to meet disposal needs. As early as 1988, it had become clear that the perceived scarcity of LLRW disposal options had prompted waste reduction activities among generators. A utility representative addressing the annual DOE conference on LLRW management noted that commercial LLRW volumes had dropped by one-third between 1984 and 1987. The nuclear utility portion of this waste stream dropped by more than 40 percent during this time, even as the number of nuclear reactors online had increased from 79 to 99. He remarked, "We used to talk about the need to conserve precious disposal capacity. Now it sometimes seems that the waste is the valuable commodity" (Farrell 1988, 16). By the early 1990s, academics were arguing that economic efficiency dictated that the United States pursue far fewer LLRW sites than Congress and the states had envisioned under the LLRWPA (Coates et al. 1994; Gershey et al. 1990).

The Rocky Mountain Compact, consisting of Colorado, New Mexico, and Nevada, concluded in 1990 that its region was not generating enough commercial LLRW to justify development of a new disposal facility. When the disposal site in Beatty, Nevada, which had been accepting just 5 percent of the nation's LLRW volume, reached capacity and closed in 1992, the compact found the Richland, Washington, site and the Northwest Regional Compact willing to accept LLRW shipments for a surcharge under a long-term contract. This relieved the Rocky Mountain Compact, and the projected host state of Colorado, from having to conduct a search for a new LLRW site. National LLRW disposal never regained its peak volume from 1980, when the LLRWPA was created, and by 1993, disposal volumes had been reduced by 75 percent.

Just as the Supreme Court was stripping the take title provision out of the LLRWPA in 1992, the South Carolina general assembly voted to allow the Barnwell LLRW facility to continue accepting waste from outside the state and outside the Southeast Compact until 1994. Governor Carroll Campbell argued that accepting out-of-region waste in the short term would provide much-needed revenue for the state budget. States could gain access to the Barnwell facility provided they demonstrated active pursuit toward the provision of disposal for their LLRW generators and paid an access fee of $220 per cubic foot on top of the disposal charges. This led many states to fall into a kind of holding pattern in which they created new siting processes, often adopting a "voluntary" approach. Some state processes, such as that in New Jersey, worked in earnest, whereas others, like those in Connecticut, Ohio, and Pennsylvania, likely were never intended to identify sites. Either way, none of these states actually advanced to identify candidate sites or progress along implementation milestones while sending their waste to Barnwell. It seemed that most states were willing to pay for a process that did not yield sites and pay hefty access fees to the Barnwell facility in order to avoid hosting a new facility.

South Carolina did close the Barnwell facility to LLRW from outside of the Southeastern Compact Region in the summer of 1994. The DOE annual report on LLRW management found that the lack of access to disposal capacity did not create "serious problems at this time or in the near future" (DOE 1995 7). A survey by the Organizations United for Responsible Low-Level Radioactive Waste Solutions reported that most commercial generators of LLRW expanded on-site storage capacity and dedicated staff to manage the waste on-site (*Nuclear News* 1996, 42). Indefinite interim storage emerged as a year-to-year solution to stalled implementation.

If a new facility were to be built, the two candidate counties that remained under serious consideration were Wake County, North Carolina, and San Bernardino, California. The Wake County siting effort met the sixth DOE milestone, which was the submission of a license application, but ran into difficulty at this point. In North Carolina, the state Department of Radiation Protection (DRP) was given regulatory authority over the license approval process. North Carolina officials began to worry that the Wake site would become not simply the first of a series of regional sites in the Southeast, but a national site. Representative George Miller, who had championed the siting process in North Carolina, explained to me in a 2003 interview the changing calculus: "Initially I was hoping we'd have a site available, because we could get our time in and get out. I was wrong because I just didn't know what the future was going to hold. Those of us that were involved in the compact process found that it was clear that the other compacts around the country weren't moving. We saw across the country that not a single site was being developed anywhere. Then we had to be concerned, well, how many next states are there going to be? And what is really going on in your own compact, and is this really a true system or not? And it went downhill from there."

According to John Runkle, an environmental lawyer involved throughout the siting process, the DRP was more rigorous in its assessment of the Wake site because of this new political calculus. "The regulatory agency took their responsibility for hundreds of years of protection seriously," he told me. "The first place that sited an LLRW would take everything for the entire country—I think that was fairly apparent to everybody."

The Chatham County Site Designation Review Committee, led by research coordinator Mary MacDowell, had been challenging the North Carolina LLRW Siting Commission on the technical suitability of the Wake site for years, to no avail. The Wake site sat atop the Triassic Basin, which all geologists working on the site had agreed was unpredictable with regard to groundwater flow. This fact did not halt the Siting Commission from reporting favorably on the characterization of the site and the environmental assessment. However, by 1995, MacDowell found the DRP receptive to the Chatham County technical objections. The DRP refused to grant a license for the Wake site. North Carolina halted the siting process and withdrew from the Southeast Compact.

North Carolina's failure to site an LLRW facility infuriated state officials in South Carolina. A spokeswoman for South Carolina governor David Beasley said, "We were the ones holding the compact system together, but the compact system is a farce when you can't compel your own members to comply. It's tantamount to having a kindergarten without a 'time-out' corner" (Smothers 1995, A10). At the same time, a year after restricting access to the Barnwell facility, the South Carolina house Ways and Means Committee began looking at opening access to Barnwell to offset property tax cuts and provide funding for education. The state general assembly successfully passed a rider to a budget bill that would reopen Barnwell for 10 years to take waste from every state *except* North Carolina. The rider raised the Barnwell access fees from $220 to $315 per cubic foot. South Carolina also withdrew from the Southeast Compact. Holmes Brown, executive director of the LLRW Forum, a group representing state siting authorities and waste generators, surmised that this development made it "likely that many states will put their planning processes on hold and conclude that Barnwell will be available as a disposal resource for some time to come" (Smothers 1995, A10).

By the end of 1995, the last candidate site in process was San Bernardino, California. The proposed site in western San Bernardino County had generated even fewer collective acts of public opposition than had the Wake County site in North Carolina. In the first six years following the public designation of San Bernardino as the candidate county (1987–1992), LLRW opponents organized just 10 collective acts of public opposition. The siting process was meeting implementation milestones by all of the prescribed dates. In fact, in 1989, the San Bernardino site had reached milestone 6, the submission of a license application, when most other states were just starting to identify candidate sites. In 1993, the California Department of Health Safety (DHS) approved the license application for the San Bernardino site. This was the first successfully licensed new LLRW site in the United States since 1970. In annual reports throughout the mid-1990s, the DOE held up the San Bernardino site as an example of successful implementation progress. By the mid-1990s, other states and regional compacts that had siting difficulties began to request permission from California to send future LLRW shipments to the proposed San Bernardino disposal facility.

In response to these developments, statewide environmental groups and elected officials began to fear that California would become a national LLRW disposal site. Normally, construction of a new disposal facility would begin immediately following license approval. However, the San Bernardino site was located on land owned by the U.S. Bureau of Land Management (BLM), not the state of California. Therefore, before construction could begin, the land had to be transferred from federal to state ownership. This final hurdle presented an opportunity for opponents to prevent successful implementation. Figure 6.3 shows a dramatic increase in collective acts of public opposition generated by statewide LLRW opposition groups in California once the DHS granted the site license in 1993.

These statewide groups included California chapters of Greenpeace, Public Citizen, and Committee to Bridge the Gap; statewide Native American tribal councils; coalitions of scientists, academics, and physicians; and a coalition of celebrities led by actor Robert Redford. These groups organized public relations

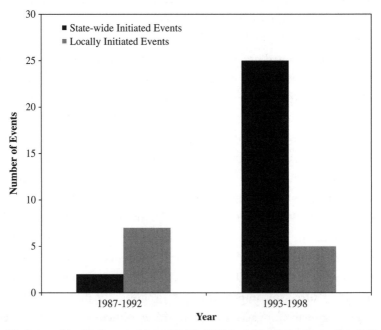

Figure 6-3. State and Locally Generated Acts of Public LLRW Opposition in San Bernardino, California

campaigns against the LLRW siting proposal, held encampments on the proposed site, and filed lawsuits claiming that the site failed to comply with the Endangered Species and National Environmental Policy Acts.

State officials, such as Controller Gray Davis and Lieutenant Governor Leo McCarthy, also began to work against the land transfer. Davis and McCarthy, two of the three members on the California State Land Commission, refused to agree to the land transfer unless Congress granted California the exclusive right to refuse LLRW importation from other states. The proposed LLRW site was an issue in the California senatorial races during the 1990s. Both Democratic senators, Barbara Boxer and Dianne Feinstein, vowed to fight the San Bernardino site. The support of these senators led to numerous time-consuming investigations and studies by the National Academy of Sciences, General Accounting Office, U.S. Geological Survey, and U.S. Fish and Wildlife Service. The California senators also had an ally in Secretary of the Interior Bruce Babbitt, who refused to transfer ownership of the BLM land. When Gray Davis won the California gubernatorial election of 1998, he officially ended the LLRW siting process.

The implementation experiences in Texas, North Carolina, and California demonstrate that the LLRW siting process was vulnerable to obstruction even in cases where local residents failed to mount significant active opposition. This vulnerability was due in large part to a changing national political environment for LLRW disposal, in which many states had ceased attempting to locate LLRW sites and the U.S. Supreme Court had removed the incentives for them to do so. In the mid-1990s, government officials in North Carolina, California, and Texas believed that the first LLRW site constructed would become a national site and thus bear a disproportionate burden of risk and liability. This perception made the LLRW siting processes in these states particularly tenuous, as they were open to the influence of state courts, legislatures, governors, and bureaucracies. Consequently, a small number of contentious acts initiated by influential governmental actors could thwart implementation.

CONCLUSIONS

When the National Governors Association and National Conference of State Legislatures successfully lobbied the federal government for a devolution of federal responsibility over commercially generated LLRW, new disposal facilities were projected to be in place within five years. Nearly 20 years after the passage of the LLRWPA, in a report titled *Low-Level Radioactive Wastes: States Are Not Developing Disposal Facilities*, the GAO said that states had collectively spent $600 million, only to have each siting process fail to establish a new site. Still, the GAO found that nearly all generators of commercial LLRW had access to waste disposal services at sites in either Richland, Washington, or Barnwell, South Carolina, and that the volume of LLRW disposed of each year continued to drop. A shortage of waste disposal availability in the United States was not predicted to occur before 2009 (GAO 1999), and no serious shortage has occurred since that time. When South Carolina governor Richard Riley led the charge to devolve LLRW disposal authority to the states, he lamented the fact that "nuclear waste stays where it is first put." If that was the case for South Carolina in 1980, it was still the case in 1999. State responsibility for the disposal of commercially generated LLRW was meaningless, as nearly all LLRW is still shipped to two national sites that were in existence before the LLRWPA.

Legal scholar Robert A. Kagan has argued that modern American government has borne significant costs of policy failure resulting from a "fundamental mismatch" between "an activist, regulatory welfare state through the legal structures of a reactive, decentralized, nonhierarchical governmental system." He laments, "Americans want government to do more, but governmental power is fragmented and mistrusted," reserving a large implementation role "for potentially recalcitrant state and local governments" (2001, 392, 396–397). The devolution approach to LLRW disposal reflects this mismatch and its fundamental characteristics—contentious, uncertain, unresolved politics. The LLRWPA simultaneously initiated a massive search for numerous new disposal facilities, while shifting authority for that search to lower levels of government with much flatter hierarchies of power. The unsuccessful effort to site new LLRW facilities revealed that the LLRWPA was poorly designed to address both the influence of local active opposition on state government actors and the power dynamics among states.

The first path of implementation is now well known: early, active, and frequent opposition thwarts implementation by winning allies within state government. Legally, states would seem to have the requisite authority to preempt, and most of the states in this study followed an implementation strategy that excluded local involvement and relied on the legal preemptive authority of states—often referred to as a top-down, command-and-control, or centralized approach to siting LLRW facilities. The siting agencies and consultants in these states did not give local citizens or governments the opportunity to choose whether they wanted their county to be considered a candidate site, did not consult with local citizens or governments, and did not inform candidate counties that they were even under consideration until after the sites were publicly announced. All of the candidate counties in the lowest quartile of implementation progress were in states that took such an approach, and all of these counties expressed high levels of active opposition.

Nearly all academics that have covered the implementation phase of the LLRWPA and LLRWPAA have critiqued this top-down implementation strategy. Michael B. Gerrard, a lawyer who represented Cortland County in its fight against an LLRW site, maintained that the top-down approach amounted to the "pre-emption of local control," which "magnifies the sense of incursion and never works in the face of determined opposition backed by the local government" (1994, 170). Kearney and Smith argued that the naming of candidate sites under this approach was a "blunt trauma for the designated communities ... after the initial shock, citizen hostility ensued" (1994, 623). The following quotes from letters to the editor in Cortland County, New York, and Rowan and Richmond Counties, in North Carolina, illustrate the hostility a top-down approach incites among local residents:

> The idea of a state-wide radioactive waste dump being forced down the throats of residents of any unwilling community in this state is totally repugnant and completely unacceptable. (*Cortland (NY) Standard* 1989)

> It seems the governor and some state officials want to shove [the LLRW facility] down the throats of the people in and around the proposed sites. ... Let's think fast and fight hard to prevent a waste dump in North Carolina. (Miller 1989)

> Regarding the two proposed "dump" sites in Richmond County, please remember "We the People" are not powerless. The government should be our servant not our lord. Fight for your rights. (Nettles 1990)

Reactions like these to state coercion fit naturally into an injustice frame, which analysis in previous chapters linked to the mobilization of active opposition.

The Connecticut implementation strategy was the most extreme example of a top-down approach. On the day the state announced the candidate sites for an LLRW facility, the spokesman for the siting agency declared that "public opposition will not be enough to deter the agency from constructing the facilities" (Hatch 1991). Yet, as the analysis in the earlier part of this chapter described, citizen opposition was fast, furious, and matched by opposition first among local officials and then among state representatives. Two state representatives penned a letter to the editor declaring that the decision "came as a shock to all members of our communities" and wrote that they were "particularly disturbed that the selection committee did not set foot on the three sites or the towns involved." The representatives vowed "to bring the committee down to earth" and concluded with the phrase "and now we fight" (Rennie and Graziani 1991). In less than a year, public opposition had indeed deterred the agency from constructing an LLRW facility.

Hanson explained that the reason local governments are often able to exert influence on the state policy process in spite of their lack of legal authority under the federal system is that "the political process makes state officials sensitive to the policy preferences of county and city governments. State legislators are elected on a local basis" (1998, 5). Lowi et al. came to a similar conclusion in their research on the politics of siting an atomic accelerator in Weston, Illinois: "County government is, after all, a unit of state government. But our studies suggest that more often than not it is a case of tail wagging dog. The power of the state of Illinois seems so often to amount to little more than the power to embrace local values and local demands" (1976, 25). The LLRW siting process supports this contention. States failed to implement the LLRWPA because they did not have the political will to preempt local opposition. Active citizen opposition begat local government opposition, which begat state representative opposition and often begat gubernatorial opposition.

In anticipation of local opposition, at least one waste industry consultant report advocated "incentives, compensation, and other magic tricks" (Visocki 1988), and others recommended "voluntary" approaches to identifying candidate counties. But even magic tricks could not avoid implementation failure, as LLRW candidate counties in the second path reveal. Both Illinois and Nebraska attempted a voluntary approach to site selection, and both offered generous compensation packages. In both cases, the definition of "community consent" is contentious. In each of these cases, the state accepted municipal government support as community consent. This caused an enormous amount of political turmoil, because the governments of the surrounding county, nearby counties, and nearby municipalities opposed the proposed site. Compensation offers for LLRW not only failed, but also seemed to incite opposition or "crowd out" any willingness that may have existed in individuals to make a contribution to the public good because the payments were perceived as improper (Frey and Oberholzer-Gee 1997). Landowners near the proposed LLRW site in Wayne County, Illinois, who were approached by IDNS with an offer of compensation, likened the situation to "a child molester luring us with candy" (*Mt. Vernon (IL) Register-News* 1988). Similarly, a resident in Clark County, Illinois, argued that "these payments give the appearance of bribing communities. ... Money cannot buy health and wealth. Wisdom and knowledge must prevail" (Cook 1990). Although implementation progress in these cases advanced farther than in the counties on the first path, local opposition eventually thwarted implementation.

Finally, local active opposition on the first two paths of implementation failure had both a direct and indirect effect on implementation across the country. Intense local active opposition quickly removed many candidate sites from consideration and cultivated champions to oppose the implementation of the LLRWPA in state government. It became clear that some states were effectively thwarting implementation of the LLRWPA. The U.S. Supreme Court removed the harshest policy tool available to force states to comply with the act, waste volumes were falling, generators were developing interim storage strategies, and existing waste sites were extending access to disposal in exchange for hefty surcharges. A perception emerged that the first newly established LLRW facility would not be a state or regional facility, but a national waste sink. This led even states that had identified candidate counties expressing relatively low levels of active opposition to take steps to halt implementation.

CHAPTER 7

PREDICTABLE DISINTEGRATION AND STABILITY: THE FRAGILE EQUILIBRIUM OF LLRW DISPOSAL POLICY

H. L. Mencken once wrote, "There is always an easy solution to every human problem—neat, plausible, and wrong" (1949, 443). But whether a solution is judged to be neat, plausible, or wrong depends on how the problem is defined. Academic accounts of the LLRW issue in the 1990s universally regarded the LLRWPA "solution" as a failure. Michael B. Gerrard, an environmental lawyer, wrote that "few laws have failed so completely" (1994, 3). Political scientists Kearney and Smith subtitled their analysis of the LLRWPA in Connecticut "Anatomy of a Failure" (1994). Economic analysts such as Coates et al. argued that the state-based system in the LLRWPA "failed to provide safe, accountable, and timely disposal of radioactive waste" (1994, 537).

The LLRWPA did fail to establish new state and regional LLRW disposal facilities. Indeed, none of the 21 LLRW candidate counties in this study ever came to host an LLRW facility. While this clearly means that the LLRWPA has been an implementation failure, in that states have not fulfilled the stated purpose of the statute, it does not necessarily mean that the outcomes of the LLRWPA as a policy of devolution to states and regional compacts have been a failure. In fact, the LLRWPA remains in force and has weathered numerous assaults, including New York governor Cuomo's pledge to "get a more intelligent federal law" (Peterson 2002, 226) and South Carolina's departure from and rejection of the Southeast Compact. In 2008, the National Governors Association reaffirmed its support for the LLRWPA with a policy position asserting that "states possess the technical and administrative capacity to manage low-level waste" (NGA 2010). The Low-Level Radioactive Waste (LLW) Forum, a group representing the state and regional siting authorities that have failed to establish new disposal facilities, as well as waste generators and the disposal industry, issued a position paper supporting the LLRWPA in 2006. The paper argues that under the LLRWPA, "disposal access exists for all classes of low-level radioactive waste from all states in the country," "commercial low-level radioactive waste is currently well regulated and managed safely," and "there is not an immediate crisis. The current national waste management system affords flexibility to make adjustments as conditions across the country change" (LLW Forum 2006, 2–3). The LLRWPA maintains such support and remains in place because it is central to an unusual and unintended waste management equilibrium that has emerged in the last decade.

EVOLUTION OF DEVOLUTION

In 1980, Congress and the states constructed the LLRWPA to address what they defined as a supply problem of limited waste disposal capacity, a technical problem of safe disposal facilities, an authority problem regarding decisionmaking over new facilities, and an equity problem among states hosting facilities. The states, spurred on by those hosting existing LLRW facilities, aggressively lobbied for an act that would both give states authority over the site selection process for new facilities and provide for a more equitable distribution of such facilities among the states. These state demands were borne out of distrust in the ability of the Atomic Energy Commission and its successor federal agencies to safely manage radioactive waste. For decades, weaknesses in the design of nuclear policy led the federal government to aggressively pursue a distributive role by promoting nuclear power, while only weakly pursuing a regulatory role. By the 1970s, regulatory neglect led to several high-profile accidents and controversies, including the near meltdown at the Three Mile Island nuclear power plant in 1979. Representative Morris K. Udall explained that the LLRWPA "put the responsibility [for LLRW disposal facilities] squarely on the States, and they want that responsibility" (*Congressional Record* 1980a). Supporters of the LLRWPA argued that devolution would ensure the creation of as many as 16 regional disposal facilities built in locations amenable to the state governments. Once amendments to the LLRWPA were passed in 1985 to provide strict timelines and milestones for state implementation, the states began identifying candidate sites to host the new facilities.

It was during this process that state siting authorities added a new implicit wrinkle to the LLRW problem definition: NIMBY, a frame capturing the fear of local opposition. Congress never considered the issue of local opposition to LLRW facilities, and the states never raised this concern while lobbying for the LLRWPA. Yet those charged with actually locating candidate sites acted on this concern in the very way they designed and implemented site selection. While siting agency officials tended to explain the site selection process in terms of technical criteria, it is clear that many of these counties were chosen not because they offered the best possible physical characteristics for hosting an LLRW facility, but instead because those making the site selection decisions deemed them least likely to oppose the construction of such a facility. Archival evidence from court documents and independent investigations in Illinois, New York, and North Carolina reveals that state siting agencies and contractors improperly elevated socioeconomic data over technical criteria such as geology in an attempt to select a site that would not generate local opposition. The candidate counties examined for this project closely fit the demographics that siting professionals of the time associated with acquiescence.

DEVOLUTION AND REVOLUTION

Ironically, those hired to identify acquiescent counties grossly underestimated the potential for active local opposition in these low-income communities and even may have helped foment the active opposition that thwarted implementation progress. The addition of demographic profiling to technical site selection was an attempt to control the "rules of the game" that determine who participates, what is to be decided on, and how the decisions will be made to ensure that candidate site selection would avoid communities with significant stores of political resources. This approach misunderstood the application of local power. Demographic profiling relies on preexisting, aggregate, static measures of such things as household income, education, rural population, civic organizations, and voting records. But power is not static—it moves. And the movement is generated by human relationships responding to political stimuli expressed and interpreted through the strategic communication of meanings by political actors. The site selection processes themselves were the political stimuli to which local opponents responded. Because the siting authorities often described site selection as an objective technical process, the revelation or suspicion of even the slightest injection of demographic or other "nontechnical" considerations into siting decisions undermined the legitimacy of the processes and enabled opponents to mark the authorities as technically incompetent, politically motivated, and fundamentally undemocratic. Opponents effectively attached a sophisticated injustice frame to siting processes across the candidate counties and offered an alternative problem definition to the LLRW issue—one that challenged the very generation of the waste, the feasibility of disposal technology at new sites, equity across communities, and democratic fairness in site selection.

The human relationships that responded to the site selection stimuli coalesced into patterns of social mechanisms. In many of these cases, opposition leaders appropriated key social sites in the community, cultivated a countywide identity of opposition, and seized on whatever political resources were available to generate effective active opposition that percolated all the way up to the state governor. The type of active opposition may have varied across these cases from disruptive to conventional political acts, but the result was the same—a cessation of the siting process. Although states had legal preemptive authority over the candidate counties they identified to host LLRW sites, they lacked the political will to act on this authority when faced with local opposition.

When New York succeeded in persuading the U.S. Supreme Court to strike down the "take title" provision of the LLRWPA amendments in

New York v. United States (1992), and South Carolina extended access to the Barnwell facility in 1995, the states faced no penalties for permanently halting the search for LLRW disposal facilities other than waste disposal surcharges. Even the few states that had been making implementation progress in the absence of significant local opposition took steps to halt site selection out of fear that successful implementation would win them a national LLRW facility.

CONVOLUTION: A NEW EQUILIBRIUM

The LLRWPA failed to address the equity issue among states—an explicit aspect of the LLRW problem definition put forth by Congress and the states. LLRW disposal facilities remain nationally centralized, rather than distributed throughout numerous new state and regional sites. The act also failed to address the implicit NIMBY aspect of the problem definition held by state siting authorities, who failed in their efforts to create new LLRW facilities in willing, or at least acquiescent, communities. But despite these failures, unintended elements of the economic and political context shaped by the LLRWPA have managed to satisfy other aspects of the LLRW problem definition put forth by Congress, the states, and even the local opponents who mobilized to thwart implementation.

In 1995, South Carolina left the Southeast LLRW Compact and reopened access to the Barnwell LLRW facility to generators nationwide for 10 years. In that same year, a private waste facility originally established to receive naturally occurring radioactive materials from uranium ore processing in Tooele County, Utah, began accepting commercially generated "lightly contaminated" radioactive wastes, limited to specific radionuclides and low concentrations of radioactivity. These two openings for LLRW disposal availability nationwide ushered in what the DOE annual report called "a period of transition in the disposal of low-level radioactive waste" (DOE 1996, viii). A new equilibrium emerged that provided just enough disposal access to ensure continued generation, while also enabling high enough disposal rates to provide an incentive that both coaxed acceptance of the waste in South Carolina and Utah and motivated generators to continue waste reduction measures.

Risks versus Rewards for South Carolina

One of the blind spots in the design of the LLRWPA was its failure to address the power of active local opposition to thwart implementation. Although states had the legal authority to impose an LLRW facility on a local community, the influence of local power within state politics made such a move unpalatable. It seemed that a willing local host was required. When state siting authorities were trying in earnest to site new LLRW facilities, they often arranged tours of Barnwell, South Carolina, for skeptical and even hostile local leaders of candidate counties to meet citizens in a willing host community. A decade before the LLRW embargo of 1979, and well before hazardous and radioactive wastes were identified as a national policy problem, it had been the Barnwell County Council that actively persuaded the South Carolina State Development Board to recruit the private firm Chem-Nuclear to develop a radioactive waste disposal facility. No opposition was voiced at the 1971 licensing hearing for the disposal facility.

The location of the site has some characteristics identified as obstacles to the mobilization of active opposition in Chapter 4 of this book, including its proximity to political borders and preexisting nuclear facilities. The Barnwell facility is in the rural town of Snelling (population 125), which is less than 15 miles from the Georgia state border. Since the early 1950s, the area has hosted the Savannah River Site (SRS), which produces materials for nuclear weapons. In the 1990s, the SRS employed about 10 percent of Barnwell's population, and the LLRW facility employed another 2 percent. Both developments also made considerable contributions to county revenue. Throughout the 1980s, local officials in Barnwell also promoted the area as a favorable site for high-level radioactive waste. When the LLRWPA implementation milestones, which included reduced waste volumes and restricted access to the Barnwell LLRW facility, approached in the early 1990s, a group of local officials and business leaders formed the Save Chem-Nuclear Task Force, and each municipal council within the county adopted a resolution in support of the facility. A 1990 public opinion survey on social and economic impacts of the LLRW facility found that nearly 75 percent of the respondents valued the economic rewards of the facility over any risks associated with its operation (English 1992, 84–87).

Unlike Barnwell County, the state of South Carolina has not maintained as consistent a love affair with the LLRW facility at Barnwell. Despite the county's willingness to continue receiving LLRW shipments from around the country, Governor Richard Riley restricted access to the site in 1979 and lobbied hard for passage of the LLRWPA—a policy designed to compel other states to find their own willing host communities. In 1983, the South Carolina legislature amended the Southeast Compact agreement for LLRW disposal to read that "in no event shall the Barnwell facility serve as the regional disposal facility after December 31, 1992," and separate state legislation required the site to close altogether after 1992. But no new "Barnwells" could be found.

Even when state siting authorities adopted voluntary site selection processes, new willing host communities proved elusive. The voluntary approach garnered significant support in the academic literature during the 1990s, based on successful hazardous waste siting processes in Alberta and Manitoba, Canada (Rabe 1994; Gerrard 1994; Munton 1996). Yet the applied voluntary LLRW siting processes failed to live up to their billing as either voluntary or successful. First, the analysis of the Illinois and Nebraska voluntary siting processes in Chapter 6 reveals that the definitions of both "host community" and "consent" are contentious. In each of these cases, the support of a municipality government was taken to satisfy host community consent, even while the citizenry at large within these municipalities—and, to an even greater extent, the citizens and governments of the surrounding county, nearby counties, and nearby municipalities—expressed vigorous opposition. As one academic critic of the voluntary approach noted, an "excessively local conception of public participation" and "placing strong boundaries around municipalities" failed to match the scope of decisionmaking with the geographic range of environmental concerns (Hunold 2002). But adopting a larger unit of review is problematic, because state siting processes have revealed that public participation is difficult to recruit until sites are selected and the technical issues associated with each site become known. And, as another academic critic notes, "there is no organization that can speak effectively for local interests" at higher levels of government (Lowry 1998, 756).

Yet it is at these higher levels of government where state and federal actors engage in consideration of the broader questions of waste management— including waste generation, volume reduction, and disposal needs. Advocates of the voluntary approach credit much of the success in Alberta and Manitoba to government engagement with potential host communities in broader decisionmaking processes (Rabe 1994). The U.S. LLRW siting attempts were less than voluntary, in that the decisions on the broader issues of waste management were off the table; the only decision under consideration was whether to host a disposal site. As an advocate of the voluntary approach, Rabe warned that this would create a perception that the siting process was "a zero-sum game ... that leaves the host a clear-cut loser, instead of part of a broader collective strategy" (1994, 156). A true voluntary process must allow candidate sites to opt out of the process at any time. But the LLRWPA was designed to compel states to create disposal sites, thus removing the possibility of a decision not to site a facility anywhere. Too often, this transformed voluntary site selection processes into a high-stakes game of "not it!," leaving the last remaining candidate site trapped as the preferred site, even if it no longer saw itself as a volunteer.

In Illinois, the pressure to establish a site led the siting authority to pursue the last remaining "voluntary" site, even in the face of unsatisfactory technical site criteria. Generous compensation packages only served to further stigmatize the siting process with the appearance of bribery for the assumption of an unfair burden. One cartoon created by an LLRW facility opponent to the voluntary process in New Jersey captured the creation of this stigma perfectly, with a picture of a fistful of cash next to the words "HOW'D YOU LIKE TO OWN A DUMP ... CHUMP?!!" (Weingart 2007, 359). The issue of environmental justice arises in response to compensation packages as well. Who besides economically needy communities are likely to accept payment in exchange for the assumption of risk? Even advocates of the voluntary approach acknowledge that it cannot succeed in situations where the notion of hosting a waste facility is equated with the assumption of an unjust burden, whether by a local community, a state within a compact, or the federal system. In the mid-1990s, it seemed that no new willing hosts could be identified, which left waste flowing to those communities already familiar with and accepting of the risks of hazardous and radioactive material.[1]

---

[1] Even in Canada, which was the greatest hope for voluntary processes, Rabe et al. have noted that a voluntary approach was "best viewed as a helpful component rather than a conclusive answer to the policy challenges posed by hazardous waste" (2000, 494). Rabe and Gunderson pointed out that in the Canadian LLRW siting experience, the only willing community was "a unique community possessing historic conversance with and economic dependence on nuclear energy and research" (2008, 212).

As state after state failed to establish new LLRW disposal options, and the Supreme Court removed the take title provision from the LLRWPA amendments but left in place the ability of existing LLRW sites to raise surcharges, the amount of revenue that could be generated from accepting LLRW from across the country consequently increased. The argument Barnwell County officials had persistently made to state office holders in South Carolina for extending waste disposal access became more and more attractive. The Barnwell facility could be a revenue generator not only for Barnwell, but for the whole of South Carolina. In 1995, a newly elected Republican governor and state house of representatives began to see Barnwell as such an opportunity. They opened the facility to all states except North Carolina and imposed a state tax of $235 per cubic foot of disposed wastes to fund higher-education grants and other education programs. When the increased fees caused generators to reduce waste volumes, the tax revenue also declined, and South Carolina created a new method for taxing disposal at Barnwell. The state required Chem-Nuclear, the operator of the facility, to pay the state a minimum of $24 million each year, regardless of the number of cubic feet of waste disposed at Barnwell. This led Chem-Nuclear to increase disposal rates. Between 1979 and 2004, the average commercial LLRW disposal rates in the United States rose from just $1 per cubic foot to more than $400 per cubic foot. Barnwell was charging over $1,500 per cubic foot for the most dangerous and difficult-to-handle LLRW (GAO 2004, 20).

But South Carolina's peace with importing LLRW remained uneasy. While the increased revenue stream may have helped state leaders look past the equity issue among states as an aspect of the LLRW problem definition, they still very much clung to the aspect of the problem definition that centered on state authority over waste management decisions. When South Carolina left the Southeast Compact in 1995, it became more vulnerable. Although the state could tax waste disposal, its ability to reject or curtail the importation of waste from out of state was much less clear under the Commerce Clause of the U.S. Constitution. Members of interstate compacts, however, were empowered by law to control, reject, or curtail waste importation. Disposal capacity at Barnwell was diminishing rapidly. In 1998, the site was thought to have capacity for another 25 years, but in 1999, site surveys revealed that it would be at full capacity in just 10 years. The state took steps to avoid this projection and ensure that it had space at Barnwell to dispose of its own waste over the long term, while preserving the option to exclude out-of-state waste.

In 2000, South Carolina entered into a new compact with New Jersey and Connecticut. This Atlantic Compact was clearly not the kind of regional compact that the LLRWPA was designed to facilitate, but it did serve to protect South Carolina's exclusionary authority over wastes accepted at Barnwell. South Carolina governor Jim Hodges proclaimed of the new compact that it "puts the environmental and energy needs of South Carolina first" (Weingart 2007, 320). Similarly, Washington State has used the Northwest Compact and a contract with the Rocky Mountain Compact to carefully circumscribe the LLRW it accepts at the Hanford facility. California and Texas, two of the states with siting processes that progressed the farthest along the DOE implementation milestones before ultimately failing, also joined nonregional compacts to shore up their authority to exclude waste importation. California joined with North Dakota, South Dakota, and Arizona to form the Southwestern Compact; and Texas joined with Vermont and Maine to form the Texas Compact.

Although the LLRWPA is just a shell of a law as far as the promotion of truly regional compacts with numerous and equitably distributed LLRW disposal sites across the states, it is an essential vehicle of protection for existing host states to exercise authority over LLRW management. For this reason, the 2010 policy statement on LLRW by the National Governors Association lauded the act because it allows states to form compacts and warned that Congress should not impose constraints on the management of those compacts (NGA 2010). Washington State warned that it would close the Richland facility if the compact system were abandoned (GAO 1999, 4). Although the LLRWPA has failed to establish new LLRW facilities, it has managed to sustain willing hosts for LLRW at existing facilities by maintaining host state authority with the compact system.

Redefining Waste in Utah

When Barnwell became a willing host community in 1971, there was no meaningful definition of low-level radioactive waste. A decade later, with passage of the LLRWPA, the poorly articulated definition of this waste stream remained, and it served as another blind spot in the policy design. Under the LLRWPA, radioactive waste was defined by its place in the nuclear power production process, rather than radioactive properties and risk to human health. LLRW serves as the residual category because it is not uranium mill tailings, spent nuclear fuel, reprocessing waste, or transuranic waste from weapons production. As a result, LLRW has heterogeneous physical properties and contains both relatively benign and highly dangerous materials.

This causes two problems for policy implementation. First, in order to be safe, disposal facilities need to accommodate the most highly dangerous and long-lived radioactive materials, even though the bulk of the waste volume does not warrant this level of protection. In contrast, the International Atomic Energy Agency (IAEA) categorization scheme sorts waste by initial radiation levels and half-lives. This scheme enables materials with similar properties, posing similar human health risks, to be managed together.

Second, the heterogeneous waste stream cultivates a public perception that the LLRW siting authorities are not competent or trustworthy. Many residents, in interviews and letters to the editor, expressed an initial understanding of LLRW as equating to a low level of radioactive risk. When they learned that this was not the basis of the definition, these residents felt as though the government was being manipulative. One Nuclear Regulatory Commission (NRC) official, Paul Lohaus, told me in a 2004 interview that this public perception is widely recognized as a problem in the regulatory community. "Once you tell people what this stuff actually is," he said, "they get angry, because they assumed 'low level' meant a low level of risk. At that point, you have lost their trust."

This scenario played itself out in Allegany County, New York, one of the candidate counties that were most actively and disruptively opposed to an LLRW facility. Ted Taylor, who became an important leader among the opposition activists, had recently retired to Allegany when it was named as a candidate site. Taylor was a physicist specializing in the miniaturization of nuclear weapons. He had worked at Los Alamos on the Manhattan Project, taught at Princeton, served as deputy director in the Defense Department for Atomic Support, and worked on the presidential commission examining the Three Mile Island accident. His life and work on nuclear fission were chronicled in John McPhee's book *The Curve of Binding Energy* (1974). Taylor recalled that when Allegany was first named as a candidate site, he was "surprised that there was so much concern expressed" in the community. He then explained his transformation into an active opponent of the LLRW facility:

> I then discovered that what was included in "low level" wastes was what I would call "high level" waste. It was an accident of definition that it came out "low level" waste. I got very upset by that. ... It was a problem of mislabeling a dangerous substance, which would then allow someone to mismanage it. The issue is really one of responsibility. ... What would have been reasonable is that they would have said, "We will redefine things. Materials like that will be dealt with in a safe way as 'high level' waste." ... The Siting Commission always mentioned the medical isotopes, but that ignores the more dangerous stuff. It has not been up front. They've ignored the most dangerous material. (Taylor 1995)

The NRC has argued that LLRW management, under the LLRWPA definition, "poses broader, more direct and ubiquitous potential risks to the health and safety than any other activity" (NRC 1997). As Paul Lohaus told me, a new statutory classification system would make it easier for regulators to "make the punishment fit the crime," meaning that disposal requirements could be more easily tailored to radioactive risk. Materials currently defined as low-level under the LLRWPA that actually pose high-level risks could then be stored separately from waste that poses less risk. And the term *low-level* would equate to the level of risk the public is being asked to bear. The NRC did adopt regulations governing LLRW disposal facilities in 1983 that recognized different radioactive properties and risks associated with the waste stream. These regulations established Class A wastes as those with the lowest concentrations of specific radionuclides. This portion of the LLRW stream makes up nearly 95 percent of waste volume, but just under 5 percent of the radioactivity. Nearly all of the LLRW generated by medical, research, and industrial sources are Class A. Class B and C wastes contain higher concentrations of the shorter-lived radionuclides, and they account for just over 5 percent of LLRW volume and about 95 percent of the radioactivity. Nearly all of the commercially generated Class B and C wastes comes from nuclear utilities. Such wastes require more stringent packaging procedures than Class A wastes, and Class C wastes must be protected from inadvertent human intrusion for at least 500 years. The regulations simply directed disposal methods at the sites; the LLRWPA still locked states and regional compacts into a search for sites that would dispose of all LLRW.

In 1995, however, a private firm in Utah inched its way into LLRW disposal by working outside of the LLRWPA. The Envirocare facility in Tooele County, Utah, began accepting only the high-volume, low-radioactivity Class A wastes, essentially leaving Barnwell, South Carolina, and Richland, Washington, to dispose of Class B and C wastes. The DOE called this development "the beginning of a more hybrid system in which efforts by the private sector to meet market demands for waste management co-exist with the government/compact processes" (DOE 1996, iv). The Envirocare facility did not emerge from a state or regional compact site selection process under the LLRWPA, yet its operation was still dependent on the authority of both the state of Utah and the Northwest Compact, of which Utah is a member. This tenuous balance between private enterprise outside of the LLRWPA and intergovernmental dynamics under the LLRWPA created a unique response to the policy blind spot of LLRW definition.

Tooele County, Utah, is indeed a willing host for LLRW disposal. But like Barnwell, South Carolina, it cannot be considered a new willing host community. During World War II, the Department of Defense built the Deseret Chemical Depot and Dugway Proving Grounds. In subsequent decades, the county hosted a magnesium-processing facility that was ranked by EPA as the nation's worst toxic polluter for many years, a number of hazardous waste facilities, and a uranium mill tailings landfill. Tooele County commissioners even created a Hazardous Industries Zoning District to "provide areas in appropriate remote locations where hazardous and radioactive wastes may be stored, treated and disposed of in a safe manner." Like Barnwell, Tooele County has even tried to attract a national disposal site for high-level radioactive waste. Tipping fees from the waste facilities in this zoning district are the largest single revenue source for the county (Shumway and Jackson 2008, 448). By 1993, the zone hosted a hazardous waste landfill, two hazardous waste incinerators, and the Envirocare facility. In 1988, Envirocare had successfully obtained a state license to use its uranium mill tailings disposal site for the disposal of other naturally occurring radioactive materials. In 1991, the company obtained another license from Utah to dispose of bulk materials, such as soil, that were "lightly contaminated" with low concentrations of specific radionuclides.

This expanding waste stream at the site did not overlap significantly with the waste accepted at Barnwell and Richland until 1995, when Utah, in conjunction with the Northwest Compact, authorized an amendment to the Envirocare license specifically including Class A LLRW. The amended license established precise limits on radionuclides, concentrations, and specifications on the physical and chemical properties of the wastes—exactly the kinds of distinctions affecting waste characteristics and risks that site selection processes operating under the LLRWPA could not make.

The Envirocare facility was not the product of a new host facility search under the LLRWPA; it was instead the incremental expansion of a private waste disposal company licensed by the state of Utah and approved by the Northwest Compact. The GAO described the development this way: "Envirocare developed the facility outside the framework of the compact act but with the acceptance of the Northwest compact." The Northwest Compact resolution on Envirocare states that "only those wastes that are approved by the state/compact of origin are provided access to the region for disposal at Envirocare" (GAO 1999, 93). And Utah state law requires approval of local government, state regulators, the governor, and the legislature to expand the classes of waste accepted at the facility.

The precise limits in the license enabled Envirocare to operate a very different kind of LLRW facility from those at Barnwell and Richland. The Envirocare facility did not require expensive remote handling equipment for waste containers, and it allowed more flexible, spatially efficient disposal techniques. Class A waste could be emptied from shipping containers, mixed with dirt, and spread out in layers in the disposal area, whereas Class B and C wastes must be packaged in permanent concrete waste forms before shipment.

Envirocare sought and obtained permission from the Northwest Compact and the state of Utah to import out-of-state and out-of-compact Class A LLRW. The facility immediately began receiving high volumes of Class A LLRW, thus allowing for more capacity at Barnwell and Richland. The GAO found in 1999 that the large volumes of LLRW received by the Envirocare facility were extending the projected capacities of Barnwell and Richland (GAO 1999, 35). In 1998, for example, compared with the Barnwell facility, the Envirocare facility disposed of more than five times the volume of LLRW. Yet the waste received at Envirocare contained only 127 curies of radiation, compared with 330,000 curies in the much smaller volume of primarily Class B and C wastes received at Barnwell (GAO 1999, 37–38). Even the members of the Atlantic Compact—New Jersey, Connecticut, and South Carolina—with preferred access to the Barnwell site, have sent more than 50 times the volume of Class A waste to Envirocare in Utah than they did to Barnwell (Vergakis 2009). The development of Envirocare *outside* of the LLRWPA, but still within the intergovernmental context established by the LLRWPA, solved a problem created by a blind spot in the law. LLRW was now being sorted according to the physical and chemical properties of the waste and risks associated with those properties.

The state of Utah provided Tooele County as a willing host for only Class A LLRW, while Barnwell continued serving most of the nation's disposal needs for Class B and C wastes. Beginning in 1998, Envirocare petitioned repeatedly for a license that would allow acceptance and out-of-state imports of the more radioactive, higher-risk classes of waste. In 2001, the company containerized waste-handling facilities to manage the more dangerous types of LLRW. However, the Utah legislature banned Class B and C LLRW from being disposed in the state, and Governor Jon Huntsman vowed to oppose any expansion, so the company withdrew the license application in 2006. That same year, Envirocare purchased Duratek, the parent company operating the Barnwell LLRW facility, and renamed itself EnergySolutions. Duratek president Robert Price said of the merger, "Duratek and Envirocare are largely complementary, we serve the same customer" (Fahys 2006). One company, under a new name, owned most of the commercial LLRW disposal capacity for all three classes of waste. The Tooele County, Utah, site was receiving 99 percent of the nation's Class A commercial LLRW, and the Barnwell, South Carolina, site was receiving 99 percent of the nation's Class B and C commercial LLRW (GAO 2008, 1).

Please Mess with Texas?

In 1996, with the Tooele County facility in Utah accepting Class A LLRW from out of state and the Barnwell facility once again open to the nation's Class A, B, and C LLRW, the DOE reported that "decisionmakers and project sponsors in some states question whether new disposal sites are needed at all." Yet the report went on to warn that "the on-again, off-again nature of access to Barnwell ... suggests to many that it is not a viable long-term solution," and that "limitations on the kinds of waste the Utah facility can accept" would eventually necessitate "new disposal capacity for the entire range of low-level radioactive wastes generated in each state or region" (DOE 1996, 5).

South Carolina maintained a love-hate relationship with the Barnwell facility. When the state abandoned the Southeast Compact in 1995, Governor David Beasley opened the Barnwell facility to out-of-state waste until 2004. The Atlantic Compact Act, which brought South Carolina together with New Jersey and Connecticut, also extended Barnwell's availability to the national LLRW stream until 2008. However, this act also prescribed a gradual reduction in the annual total waste volume received at Barnwell. EnergySolutions, the Barnwell facility's new operating company, began aggressively lobbying the South Carolina general assembly to keep the site open to the nation's waste. The company's 10 lobbyists succeeded in getting the South Carolina house to introduce a bill extending nationwide access to its facility until 2023. The bill was defeated in 2007, however, largely on the grounds that the remaining disposal capacity at the site should be reserved for South Carolina—particularly for the eventual decommissioning of the state's nuclear power generators. The facility was more than 90 percent full. Of the remaining 1 million cubic feet of space, 80 percent was already committed to compact members New Jersey and Connecticut. While local Barnwell officials, who stood to lose $2 million in annual revenue from Barnwell's restricted access, railed against the votes in the house and senate, state officials were willing to give up the $12 million in state revenue and prepare Barnwell for closure. State Senator John Courson described his vote against the facility this way: "This is a loose snake slinking around for the past decade. We intend to cut its head off" (Peterson 2007).

In June 2008, they finally did. The year prior, South Carolina raised the rates one last time, with the biggest rate hike to date, increasing annual state revenue from the site by nearly $1 million. EnergySolutions CEO Steve Creamer reflected on the company's experience with the Barnwell facility: "We didn't just get kicked out of South Carolina, we got brutalized and kicked out of South Carolina" (Wilder 2008). As capacity dwindled at the Barnwell facility, South Carolina's willingness to host a national LLRW disposal facility eventually expired. Will the closure of Barnwell upset the unusual equilibrium that has developed over the past 15 years? The existing LLRW facilities had been providing just enough disposal capacity to allow continued generation, while keeping disposal availability scarce enough to enable host states to levy hefty surcharges and motivate generators to take significant reduction measures.

What South Carolina ultimately saw as a problem, Texas began to see as an opportunity, as outlined below. Currently, the Utah facility is receiving 99 percent of the commercially generated Class A LLRW nationwide. The Richland, Washington, facility is receiving Class A, B, and C wastes for the 11 states in the Northwest and Rocky Mountain Compacts. The Barnwell, South Carolina, facility receives Class B and C wastes only from the 3 states in the Atlantic Compact. This leaves LLRW generators in 36 states with no place to dispose of Class B and C wastes. Generators will continue to minimize waste, process it into safer forms, and store it on-site until disposal options become available. A survey of generators in Pennsylvania found that most had estimated their on-site disposal capacity at 20 years (Lenton 2008). The Nuclear Energy Institute has indicated that the current situation is "not a crisis" because generators have planned for interim storage. Yet on-site storage is more costly for most generators than even the hefty disposal fees were at Barnwell (GAO 2004). If a new willing host for Class B and C LLRW emerged, it would have plenty of customers willing to pay a premium.

A decade ago, Texas would have seemed like an unlikely candidate to serve as the next willing host for LLRW—it had already failed to implement the Hudspeth County site under the LLRWPA, and in 1999, the Texas LLRW Authority was abolished. The GAO found that "prospects for significant progress" in Texas were "uncertain" at best (GAO 1999, 75). In 2002, Maine withdrew from the Texas Compact, having lost faith in the state's ability or willingness to successfully establish a new LLRW facility. This left Vermont and Texas as the only members of the Texas Compact. In 2003, a new Republican majority in the Texas house joined the Republican senate and governor to adopt a different approach. Instead of operating yet another state search for an LLRW facility site, the Texas legislature simply authorized privatized radioactive waste disposal in Texas. The company Waste Control Specialists (WCS) submitted a license for an LLRW facility in Andrews County to the Texas Commission on Environmental Quality (TCEQ) in 2004. WCS, whose primary shareholder, Richard Simmons, is one of the largest contributors to national and state Republican campaigns, lobbied hard for the new legislation and helped craft it. In the words of WCS president Rod Baltzer, "We just had to get the state law changed" (Wilder 2008).

The new Texas law was perhaps inspired by the success of Utah's development of a private disposal facility tenuously balanced outside of the LLRWPA, yet still subject to state and compact authority. Like Tooele County in Utah, Andrews County, Texas, has created "enterprise zones" specifically designed to attract waste industries. In one 1,338-acre site on the New Mexico state border, Andrews County hosts a hazardous waste landfill that accepts polychlorinated biphenyl (PCB) sludge from the dredging of the Hudson River, a radioactive waste treatment plant, and a mixed-waste treatment facility for substances both hazardous and radioactive—all owned by WCS. The county also is actively recruiting developers to build a gas-cooled nuclear reactor in the enterprise zone, with express plans of making the area a "nuclear corridor." Just across the border in New Mexico, the Chihuahuan Desert hosts a DOE disposal facility for transuranic waste from nuclear weapons production and a uranium enrichment facility. Like the Barnwell site, the Andrews County, Texas, LLRW site is on a border. In fact, Eunice, New Mexico, is the town closest to the site. And like Barnwell and Tooele, Andrews County tried unsuccessfully to attract the national repository for high-level radioactive waste now under contested development at Yucca Mountain, Nevada.

Andrews is home to a community that supports, and to a large extent depends on, the waste industry. As WCS president Rodney Balzer has boasted, "The community is welcoming it—they're encouraging us. They're looking forward to the employment." In Hudspeth County, south of Andrews, the first Texas LLRW siting attempt was defeated in part because of opposition from the county judge. That will not happen in Andrews County, where Judge Richard Dolgener has described the LLRW facility as "good for the county because it will expand the revenue stream already generated by WCS waste operations" (Folsom 2008). Dolgener also suggested radioactive waste disposal as a way to diversify the local economy: "We're trying to get off the oil tit because the bust will come one day" (Wilder 2009). The few opponents to an LLRW facility in Andrews were left intimidated by aggressive, well-orchestrated, and overwhelming displays of public support. At one public hearing, when an audience of hundreds was asked to stand if they opposed the license, only one person, the leader of the opposition group AWARE, stood up. But even this minimal opposition soon fizzled out: AWARE had a regular membership of approximately 15 people, half of whom turned out to be WCS supporters, and the group voted to disband (Wilder 2009).

Even in the face of serious questions regarding the technical suitability of the disposal site, the licensing process went forward. Three TCEQ scientists resigned their positions in protest during the process. One hydrogeologist said, "Agency management ignored my conclusions and those of other professional staff, and instead promoted issuance of the licenses." An engineer deemed the application "very deficient" because "the geology of the site is unsuitable for containment of radioactive waste for thousands of years" (Wilder 2008). He and the other resigning staff found that the water table was as close as 14 feet, which made it highly likely that water would seep into the disposal as annual rainfall increased with climate change (Elliott 2008). Yet hundreds of people from Andrews County donned green T-shirts reading, "WE SUPPORT WCS," and traveled to Austin, Texas, to express support for the LLRW facility at TCEQ meetings on the draft license and final license approval. The supportive crowds included the mayors, city managers, county commissioners, and school board members of both Andrews County, Texas, and Eunice, New Mexico. The Texas Sierra Club was unable to find a single local resident to file for contested case hearings on the license.

The license for the WCS LLRW facility in Andrews County was approved in January 2009, and as of this writing, construction had begun in the fall of 2010 and operations were scheduled to commence in 2011. Like the site operated by EnergySolutions in Utah, the WCS site is working with the state and compact authorities to negotiate access for disposal. WSC has expressed its desire to dispose of commercial LLRW of all classes from states besides Texas and Vermont. After receiving the site license, WSC vice president William Dornsife painted the facility as the next national facility: "We believe flexible import provisions would go very far toward resolving the nation's challenges ... now that the Barnwell facility no longer allows nationwide access for disposal of these wastes." Company spokesman Chuck McDonald added that accepting waste from outside the compact would generate considerable revenue for Texas (*Bennington (VT) Banner* 2009). The Texas Compact Commission can, and likely will, contract with other states and compacts to provide national disposal for all classes of LLRW at a higher price than that charged to generators in

Texas and Vermont. Thus, by following Utah's lead and enabling the development of a private LLRW disposal facility at an already existing site in an already welcoming community, Texas will likely take Barnwell's place in the LLRW disposal equilibrium—providing an expensive, but available, final resting place for the nation's Class B and C LLRW.

PREDICTABLE DISINTEGRATION AND STABILITY

Radioactive isotopes disintegrate in a predictable way, forming other isotopes and eventually reaching stability. Could this be a metaphor for U.S. policy on low-level radioactive waste disposal? As a system for creating numerous new LLRW disposal sites in an equitable distribution among states, the LLRWPA disintegrated. To what extent was the disintegration predictable? Will the unusual equilibrium that has resulted in the wake of this disintegration—which is clearly a different outcome than the law was intended to produce—achieve lasting stability?

Predictable Disintegration?

In 1955, Albert Einstein was asked why humans could discover atomic power but not the means to control it. He replied, "Because politics is more difficult than physics." This bit of wisdom should caution against stretching the metaphor above too far in the direction of predictability. Still, some aspects of the policy history shaping disintegration of the LLRWPA could have been predicted. The most significant blind spots in the design of the LLRWPA were the vague "catchall" definition of the waste stream, the failure to account for the power of active local opposition to new LLRW sites, the failure to guard against the politics of LLRW facility avoidance among states, and the gross overestimation of future LLRW volumes.

In some ways, the policy context for LLRWPA passage in 1980 should have raised awareness on these issues. By this time, it was clear that the very design and implementation of civilian nuclear policy in the United States had favored the promotion of nuclear power plants at the expense of effective regulations in ways that led to highly visible environmental and safety problems.

Even authorities such as former Atomic Energy Commission chair David Lilienthal had publicly recognized both the lack of regulatory attention and lack of waste management strategy as problems. Distrust of the federal government on the issue of nuclear power and waste was expressed not only by state governments demanding greater regulatory authority, but also by the general public. Public opinion polls reveal that opposition to nuclear power began rising in 1974, and by 1980, opposition outnumbered support by two to one (Rankin et al. 1984). Baumgartner and Jones use this shifting public tide on nuclear power as an example of punctuated equilibrium—a process whereby significant changes in the popular political image or definition attached to an issue lead to dramatic changes that "alter forever the prevailing arrangements in a policy system" (1993, 10). Clearly the equilibrium of unquestioning public support of nuclear power and underregulated nuclear waste management had already been punctuated in 1980.

The vague definition of LLRW codified by Congress in 1980 created a waste stream heterogeneous in terms of radioactivity, risk, and even source. The LLRWPA lumped high-volume, high-radioactivity wastes from nuclear power production together with relatively short-lived, low-radioactivity, low-volume wastes from medicine, research, and industry. Such a waste stream inspired opposition in local communities already primed by growing opposition to nuclear power. One wonders whether siting processes for strictly medical waste or strictly Class A LLRW would have gained more acceptance. But the issue of local acceptance or opposition did not seem to garner any attention in Congress, even though local activism in response to hazardous waste was one of the biggest national news stories in the years immediately preceding passage of the LLRWPA. Just as Congress was passing the LLRWPA, activism among local residents in Niagara Falls, New York, was succeeding in forcing initially reluctant local, state, and federal governments to pay to arrange for neighborhood relocation and cleanup of the Love Canal contamination. The LLRWPA was born out of a state distrust of federal radioactive waste management, but it would founder in the wake of local distrust of state waste management efforts.

Proponents of devolution in the U.S. federal system argue that states are better suited to implement policy, because they are more knowledgeable of and closer to local concerns. In the case of LLRWPA implementation, state siting authorities certainly seemed knowledgeable of the potential for active local opposition, but they often acted in ways that only incited the kind of opposition they feared. In instances where siting authorities strayed from technical criteria and selected candidate sites according to factors thought to be associated with acquiescent communities, they were perceived as rigging the rules of the game to select communities with few political resources. Activists in many of these communities were able to frame the siting process as an injustice and motivate effective active opposition—demonstrating that power flows not only from preexisting political resources and the rules of the game, but also from the dominant meaning ascribed to public issues and the dynamics of social mechanisms in a given political context. In the end, states were in fact responsive to local concerns, but not in ways that benefited the devolutionary approach of the LLRWPA. As opposition bubbled up from local communities to state office holders, the states undermined the intended interstate equity of the LLRWPA. It was no mystery that new LLRW disposal sites were "public bads" that states would seek to avoid, and the LLRWPA and subsequent amendments sought to provide incentives to correct for this. But it would have been difficult to predict that the U.S. Supreme Court would eventually choose the LLRW issue as a platform for the reassertion of states' rights and remove the strongest incentive in the LLRWPA preventing implementation avoidance by the states.

In the 30 years that states have spent avoiding implementation of the LLRWPA, LLRW volume has declined dramatically. Congress designed the LLRWPA with a vision of expanding radioactive waste disposal needs and sought to provide a virtually limitless supply of disposal capacity distributed evenly across the country. It failed to consider the predictable constraint that public opposition to nuclear power would have on the nuclear industry, as well as the predictable innovations in waste reduction strategies that would stem from limited LLRW disposal options. The policy history reveals that it was clear almost immediately that there was no need for a law prescribing so many new disposal sites.

Stability?

And yet the LLRWPA persists, and the authority it accords to the states and compacts for LLRW disposal holds together an equilibrium providing just enough disposal capacity to enable continued waste generation, while also motivating generators to reduce waste and enabling high enough disposal rates to coax a few already contaminated sites to accept the nation's waste. The question of whether the LLRWPA has been a successful policy depends on just how the problem of LLRW is defined. If the definition of policy is not limited to legislation, but also includes the multitude of authoritative decisions that govern public life, then LLRW policy is not only the LLRWPA legislation, but also the myriad decisions by intergovernmental actors that have conspired to maintain the current balance between waste generation and disposal capacity. The power of active local opposition and its relationship to the interstate dynamics of waste avoidance has prevented the creation of a multitude of new LLRW sites equitably distributed among the states—a key aspect of the problem definition put forth by the states in 1980 and reflected by Congress in the design of the LLRWPA. Particularly from the perspective of the states hosting existing LLRW facilities in 1980, this would surely mark the policy as a failure.

Yet for the local opponents and antinuclear groups who organized to thwart site selection, the LLRW problem was defined in a way that challenged the very generation and definition of the waste stream, as well as the technical feasibility and equitability of selecting new and uncontaminated "green field" sites without the participation of the affected communities. From this perspective, the persistence of constricted LLRW disposal capacity for the past 30 years has achieved some successful outcomes. The volume of waste production has fallen significantly, and a lack of cheap and plentiful radioactive waste disposal capacity, including LLRW, has surely played a role in preventing the expansion of nuclear power generation in the United States.[2] The Tooele County, Utah, facility's accepting only Class A LLRW essentially redefined the waste stream by radioactivity, risk, and source. Local opponents and antinuclear groups would count as a success the fact that not a single new site has been developed to host an LLRW facility. Yet from this perspective, policy success on the LLRW issue has not yet been fully realized. The position of the antinuclear group Nuclear

---

[2] The lack of cheap and plentiful commercial radioactive waste disposal capacity has, however, dramatically increased the cost of decommissioning existing nuclear power plants. Interestingly, this lack has not hampered the cleanup of federal government sites contaminated with radioactive wastes, which have largely been handled on-site or at Department of Energy facilities (GAO 2004, 17).

Information and Resource Service asserts, "We should stop making more waste" and "cease the use of nuclear power" (NIRS 2009).

Limited disposal access has constricted but not extinguished LLRW from nuclear power. Congress, the states, and the nuclear industry designed the LLRWPA to provide adequate disposal capacity, and these actors continue to support the policy because it has been successful in providing disposal—but not at fresh sites. Even in the wake of the one-year Barnwell closure in 1994, the Nuclear Energy Institute (NEI), speaking for the nuclear industry, argued that the LLRWPA and the disposal situation were "sound" and that "states are best suited to manage low-level radioactive waste materials" (DOE 1995, xii). The industry fears of crises resulting from a lack of LLRW disposal capacity, expressed before passage of the LLRWPA in 1979, have never been realized. State siting authorities of the 1980s and early 1990s added the NIMBY frame to the problem definition of the LLRW issue, fearing that local opposition would cause a crisis of disposal capacity. Indeed no new "backyards" ever stepped forward to host the waste, but just enough willing hosts with already contaminated backyards have stepped forward. These have included existing LLRW facilities such as Barnwell, South Carolina, which extended disposal access 15 years beyond the deadlines written into the LLRWPA, and Richland, Washington, which has continued to accept LLRW from the Northwest and Rocky Mountain Compacts, as well as new facilities built at existing hazardous waste sites like Tooele County, Utah, and Andrews County, Texas. These states are willing to host the sites because of the revenue they can gain from serving as the limited disposal option for the nation's LLRW producers and the control they maintain over disposal location, interstate acceptance, and regulation under the LLRWPA. They provide just enough LLRW disposal capacity for the continuation of medical, research, industrial, and nuclear power applications.

If the question of policy success is a mixed bag dependent on diverse problem definitions, it may be more useful to ask whether the current LLRW policy is stable, and what the likelihood for policy change in this area might be. Future developments could, of course, upset this stability and even punctuate the political equilibrium on nuclear power and waste disposal. The WCS LLRW facility in Texas could fail to open, or open only to waste shipments from Vermont and Texas, thus prolonging the costly interim on-site storage strategies most generators have been forced to employ since Barnwell's closure in 2008. This would tip the equilibrium further in the direction of waste minimization, volume reduction, and limitations on waste-generating activities.

If the facility in Texas fails to open, Utah and Washington, the remaining states hosting LLRW sites and both members of the Northwest Compact, have shown no willingness to expand disposal capacity. The state of Utah has worked diligently to prevent EnergySolutions from accepting the nation's LLRW at any levels higher than Class A. After the Barnwell closure in 2008, Bill Sinclair, deputy director of Utah's Department of Environmental Quality, told nuclear waste handlers and regulators at the RadWaste Summit in Las Vegas, "Don't put all your eggs in one basket," and warned that Utah was "growing wary of being known as a national treasure because of the EnergySolutions site" (Fahys 2008). In Washington, voters passed an initiative that would bar waste shipments to the Hanford site until the DOE cleans up the larger nuclear reservation on the site. The federal court struck down the measure, but the fact that voters had passed it by a two-to-one margin demonstrates a significant lack of support for Washington's host state status. The Northwest Compact has made no indication that would entertain the possibility of accepting LLRW from outside the Northwest and Rocky Mountain Compacts at the Hanford site.

Yet based on the success of WCS influence on both the design of the Texas legislation enabling private LLRW disposal and the licensing approval—even amid controversy and regulatory staff resignations over technical criteria—the Texas site seems likely to open to the nation's LLRW generators. Despite some outstanding mineral rights for the facility site and a bill in the Texas legislature proposing a ban on waste importation from outside Texas and Vermont, the WCS facility seems poised to accept all classes of the nation's LLRW. WCS president Rod Baltzer reassured an audience of waste generators and management specialists at the 2008 RadWaste conference in Phoenix: "There is a problem in the nation because there's nowhere to ship this waste, so if we can provide a solution to a national problem, we'd be interested" (Wilder 2008). In fact, WCS has already begun accepting Class B and C out-of-state wastes from a Tennessee processing facility that used to prepare waste shipments from across the country for disposal at Barnwell. The TCEQ has not objected to the company's storing these shipments indefinitely at its existing waste-processing facilities on-site, pending the opening of the permanent LLRW disposal facility. Nor has the TCEQ objected to the construction of the disposal facility expanding to more than twice the capacity necessary for Vermont and Texas LLRW disposal needs over the next 40 years.

Consequently, it is somewhat more likely that the *opening* of the WCS LLRW facility upsets the equilibrium. The large capacity of the facility and willingness of WCS to fill it quickly could expand disposal capacity—even providing a surplus—in such a way that tips the balance away from waste minimization and in the direction of expanding waste generation activities. Andrews County, which has visions of positioning itself as a "nuclear corridor" and has even developed proposals to attract a new nuclear power plant and high-level radioactive waste disposal, could become a key piece of a nuclear "renaissance" in the United States. Despite the problems nuclear power has experienced with the American public over the past 30 years, newly available waste disposal capacity in a willing county in a willing state could enable a new, more positive policy image that positions nuclear power as an "alternative energy" solution to security and climate change concerns associated with fossil fuels. This reverse punctuated equilibrium would certainly satisfy the nuclear industry. However, it is important to note that the state of Texas and the Texas Compact Commission still have an express interest in exercising authority over waste acceptance at the new WCS facility. As the commission's chairman, Michael Ford, has warned, "There is some concern that it will become a free-for-all, an open season . . . that is not what we're charged to be doing" (Wilder 2009).

The maintenance of host state and compact authority over waste management has been central to the commercial LLRW disposal equilibrium over the past three decades. A recent U.S. District Court decision concerning the EnergySolutions facility in Utah significantly weakens this authority. When EnergySolutions failed to persuade the state of South Carolina to keep the Barnwell facility open to the national LLRW stream, and failed to persuade Utah and the Northwest Compact to broaden the acceptable waste stream at its Tooele County facility to include Class B and C wastes, it began exploring international waste opportunities. The company filed an application with the NRC in early 2008 to import 20,000 tons of LLRW from decommissioned nuclear reactors in Italy for disposal at its Utah facility. The company proposed sharing half of its profits from the foreign waste disposal up to $1.5 billion over

10 years. When Utah opposed the plan and the Northwest Compact asserted its authority to ban the importation of foreign waste, EnergySolutions filed suit in the federal court, arguing that its facility was not part of the compact system under the LLRWPA and that the commerce clause of the U.S. Constitution prevented import bans. In May 2009, U.S. District Court judge Ted Stewart ruled in EnergySolutions' favor—finding that despite the 18-year agreement of operation among the Northwest Compact, Utah, and the privately owned and operated LLRW facility, the compact has no authority to limit the flow of radioactive waste to the Tooele County facility, because it is not in fact a regional facility, and the LLRWPA made no provisions for the development of private LLRW facilities. The court warned of the potential for abuse if private disposal facilities were left at the whims of the compacts. The ruling is subject to appeal, and U.S. House and Senate bills that would ban foreign imports of radioactive waste are under consideration.

However, if this court's interpretation of the privately developed facility in Utah becomes precedent, it will certainly alter the dynamics of LLRW management. Representation for the compacts in the case described this interpretation as "obliterating the compact system" (Fahys 2009). If the compact system no longer holds authority to regulate the acceptance of waste at privately developed facilities like those in Utah and Texas, the capacity for disposal in terms of both volume and class of radioactivity could expand greatly. The two companies operating the facilities in these states are direct competitors, and the price to generators for disposal would likely fall—once again tipping the balance away from waste minimization and in the direction of expanding waste generation activities. This court interpretation of the private facilities would essentially realize an alternative to the LLRWPA put forth in a 1999 GAO report. The alternative called for a repeal of the act to enable private industry to develop and operate disposal facilities in response to market conditions—freeing private waste facilities from state control over operations. The GAO was careful to note, however, that individual states would still retain licensing authority. It is highly likely that should the court's interpretation stand, Utah would cease to be a willing host and would eventually use its licensing authority to close the EnergySolutions facility. The state of Washington has also signaled that it would decline to renew the lease on its LLRW disposal site if the compact system were weakened or eliminated. This would leave the WCS facility in Andrews County, Texas, with a monopoly on commercial LLRW disposal.

Another alternative outlined by the GAO (1999) and endorsed by the Health Physics Society (2005) would have the federal government reassert itself by giving the DOE responsibility over commercial LLRW disposal. The DOE has both existing LLRW disposal sites with large unused capacities and numerous "legacy sites" that some have argued would be well suited for disposal. These latter sites include facilities that were designed for nuclear fuel reprocessing, weapons production, and weapons testing. Some of these sites are closed or repurposed military sites.[3] They often are remotely located, expansive, and possess an extensive security infrastructure. In many cases, it is not possible to fully restore these sites to clean "green field" status, and they will require constant stewardship for hundreds of years or more. A group of retired DOE and NRC employees has recommended that productive uses be found for these sites to ensure that appropriate stewardship of the sites continues (Bell et al. 2004). This alternative would greatly expand disposal capacity and greatly reduce the cost of disposal for generators, as the federal government would be essentially subsidizing the waste component of nuclear production processes— once again tipping the equilibrium in the direction of increased radioactive waste generation activities. However, Washington and Nevada, which host many of these sites, have expressed vigorous opposition to this alternative.

In addition, the DOE's inability to successfully site a high-level radioactive waste facility on federal land at the Yucca Mountain site in Nevada does not bode well for this plan and reveals a contrast between the weaknesses of a centralized approach and the advantages and unintended benefits of state-compact authority of the type currently provided under the LLRWPA, which boosters of the Andrews County, Texas, "nuclear corridor" find attractive. The Yucca Mountain site selection epitomized a centralized approach, with the federal government wielding authority over a state. The selection was not the result of a national search for sites with suitable technical characteristics. It was, instead, a political decision. In 1982, the Nuclear Waste Policy Act had laid out

---

[3] Interestingly, the successful closure of military bases is one example of the federal government acting effectively in the face of local opposition. Mayer (1995) has explored factors associated with this implementation success.

a technical site selection procedure for an eastern and a western high-level radioactive waste repository. In 1986, amid concern that sites in Wisconsin, Georgia, and North Carolina could threaten vulnerable Republican congressional seats, President Ronald Reagan suspended the site selection process and announced that the first and only repository would be built in Texas, Nevada, or Washington. Within a year, Speaker of the House Jim Wright from Texas and House Majority Leader Tom Foley of Washington championed the Nuclear Waste Policy Amendments Act, which permanently halted the national site selection process and named Yucca Mountain, Nevada, as the one and only candidate site for the nation's high-level radioactive waste.

In a framing effort that mirrors the application of injustice frames in the LLRW candidate counties, Nevada's congressional delegation and state government dubbed this amendment the "screw Nevada plan" and have worked tirelessly to prevent the Yucca Mountain site from opening. While the federal government spent more than 20 years and $10 billion characterizing the Yucca Mountain site, Nevada accumulated power in Congress and realized pressure in the Electoral College. By 1999, Nevada senator Harry Reid was Democratic Party whip, and in 2006, he became Senate majority leader. Reid has proved a formidable opponent to the site. He blocked 175 of President George W. Bush's nominations for various positions, until Bush finally agreed to appoint a member of Reid's staff to the NRC body licensing the site. Reid also managed to slow implementation by slashing the budget for Yucca Mountain by nearly two-thirds. In the 2008 presidential election, candidate Barack Obama promised Nevada he would oppose the Yucca Mountain site. In 2009, President Obama sent a budget to Congress that defunded the Yucca Mountain project and pronounced that the country would have to think of another disposal option for its nuclear waste.

The Yucca Mountain failure demonstrates the power of local and state opposition in U.S. intergovernmental relations—even under a centralized approach in which a single site was designated by Congress and administered by a federal agency on federal land. This case should caution decisionmakers from scrapping the LLRWPA and attempting a Yucca-type effort for LLRW. In contrast, what the remaining skeleton of the LLRWPA provides, especially in the absence of the imagined two dozen regionally distributed disposal sites, is host state authority under the compact system to control the import of radioactive waste and dictate the taxes and fees that will be collected for the provision of scarce disposal access. This is why the LLRW facility appeals to local boosters in Andrews County, Texas, *as well as* the Texas state officials and congressional delegation. This is a rare intergovernmental concatenation of support that has taken shape under a tenuous policy equilibrium.

Any future that emerges for the management of commercially generated LLRW in the United States will be layered on more than three decades of past intergovernmental political development in this policy area. It is a policy area that has artificially circumscribed the waste issue and failed to directly engage the full system of production, waste generation, and disposal. As one academic observer has noted, the U.S. civilian nuclear policy "has reacted to waste generation" rather than carefully examining overall waste management needs (Rabe 1994, 153). The current disposal equilibrium that has emerged under the LLRWPA is subject to a prolonged standoff between political supporters and opponents of civilian nuclear power—an issue the country has not explicitly addressed since Congress cleared the way for the promotion of civilian nuclear power in the 1950s. Instead, nuclear power generation at existing plants continues, as decommissioning is forestalled and any future expansion with new sites remains uncertain. Nuclear supporters position the industry as a carbon-free solution to the global-warming problem and promise a new generation of nuclear power plants. There is evidence that public opinion on nuclear power has shifted, with a March 2009 Gallup poll showing 59 percent of Americans favoring nuclear power as an energy alternative (Aeppel 2009). The NRC is now considering 17 license applications involving 26 new reactors.

Yet there is no evidence that public opinion has shifted on the acceptance of nuclear waste, and disposal for all types of radioactive waste, including LLRW, remains tightly constrained. Nuclear opponents rely on limited waste disposal capacity to serve as a brake on the expansion of nuclear power. Political actors at both the local and state levels have successfully applied this brake many times to thwart the creation of new disposal facilities at new sites. How long this standoff persists will depend to some degree on the future volumes and types of waste accepted at the LLRW facilities at Tooele County, Utah, and Andrews County, Texas, as well as the authority those states are able to maintain over the waste stream and the revenue that waste stream can provide.

APPENDIX

This appendix is provided to clarify assumptions, methods, and results for quantitative analysis of mobilization of opposition to LLRW site proposals, disruption as a tactic of opposition, and implementation of facility siting.

THE FACILITY SITING HYPOTHESIS: LIMITATIONS IN THE QUANTITATIVE ANALYSIS

Demographic profiling during site selection did not enable siting authorities to avoid opposition, but could key demographic factors explain the variations in levels of active opposition? Table A.1 summarizes Poisson regression on the number of collective acts of opposition to the LLRW facility in the 100 days following the first opposition event in each candidate county. The dependent variable derives from my tracking of all collective acts of public opposition to the LLRW facility in the local daily newspaper nearest to each site, from the day the candidate site was announced in the newspaper to the day the siting process effectively ended for each site. This measure controls for the different durations of siting processes across these cases by confining the event count to 100 days. It also helps control for different rates of initial organization across the candidate counties by beginning the event count with the first collective event of public opposition in each case.

The four independent variables in this analysis represent four factors common across waste industry guidance of the era: (1) the political leanings of area politicians, measured as the League of Conservation Voter (LCV) score of each county's representative in the U.S. House of Representatives in the year preceding the siting process;[1] (2) the rural composition of the area measured with 1990 U.S. Census data on the proportion of the population classified as "rural"; (3) the median household income measured with 1990 census data; and (4) the educational attainment of the local population measured as the proportion of the population 25 years or older with an educational attainment of high school graduation or less in the 1990 census data.

According to the facility siting hypothesis, high activism would be associated with relatively high LCV scores, low rural populations, a high median household income, and low proportions of the population with educational attainment of high school graduation or less. However, Poisson regression of the count data produces results that do not square well with the hypothesis (see Table A.1).[2]

The signs for both the rural population and educational attainment variables are positive and statistically significant at the 0.05 level. This means that the

---

[1] High scores on the LCV scorecard indicate support for the environmental movement's agenda on a broad range of issues. LCV scores are available at the website www.lcv.org. Although the representatives' districts do not perfectly match the boundaries of the candidate counties, they did provide the best available measure for this political issue, because environmental issues do not perfectly track voter preferences for party or ideological identification, and measurements of environmental support for elected officials at state and local levels were unavailable across these cases.

[2] Poisson regression is a method used to evaluate count data. The compilation of collective public acts of opposition considered here is counts of events. Count data violate assumed normal distribution of ordinary least squares regression, following instead the Poisson distribution (King, 1988).

Variable	Estimates (Standard error)	$P > /z/$
Constant	−2.166541 (0.9463258)	0.022
Rural population	0.9541994 (0.3178647)	0.003
Educational attainment	3.391906 (1.069326)	0.002
Income	0.0000886 (0.000012)	0.000
LCV score	−0.0049935 (0.2038753)	0.980
$N = 21$ Chi square Pseudo $R^2 = 0.1973$	61.51	0.000

Table A-1. Poisson Regression on the Number of Collective Acts of Opposition to the LLRW Facility in the 100 Days Following the First Opposition Event in Each Candidate County (Model 2: LULU Siting Literature)

more rural and less educated the county population, the more actively opposed it was during the LLRW siting process. These results do not match the expectations of the siting industry recommendations. The median household income is also significant. It displays a positive relationship with activism, as the industry experts expected, but the standard error is close to zero. The LCV score is insignificant. Overall, the model leaves much of the variation unexplained, with a pseudo R^2 of just 0.1973.

CLASSIC SOCIAL MOVEMENT HYPOTHESIS: LIMITATIONS IN THE QUANTITATIVE ANALYSIS

The classic social movement agenda puts forth three explanatory concepts for mobilization: (1) political opportunity structure, defined as access for movement participants and the presence of allies in power (McAdam, 1996; Tarrow, 1998) and measured here as LCV scores for each county's U.S. representative in the year preceding the siting process; (2) mobilizing structures, defined as "embedded social networks and connective structures" (Tarrow, 1998, 23) that link activists to "other groups" and provide "external support"

(McCarthy and Zald, 1987, 16), and measured here as people per civic organization in each county;[3] and (3) collective action frames, defined as the strategic process of ascribing meaning to movement goals and activity (Snow et al. 1986), and measured here as the proportion of days during the siting process that an injustice frame message appeared on the editorial page for each county.[4] The classic social movement hypothesis holds that (a) the higher the LCV scores, the better the political opportunity for the county facing the LLRW site, and thus the greater the activism against the site proposal; (b) the more people per civic organization, the less active the county will be because of the lack of mobilizing structures available to activists in the community; and (c) the higher the ratio of days during the siting process with an injustice frame message on the editorial page, the more collective acts of public opposition in the county.

Table A.2 displays the results of Poisson regression evaluating the impact of the LCV score, the ratio of population to civic organizations, and the prevalence of an injustice frame on activism in counties facing an LLRW facility proposal. Only two of the three variables from the classic social movement agenda are significant at the 0.05 level, and just one is significant in the expected direction. Overall, this model explains a modest amount of active opposition, with a pseudo R^2 of just 0.31.

The LCV score is significant. However, the sign of the relationship is negative. In accordance with the classic social movement agenda, I

---

[3] The concept of mobilizing structure is closely related to the concept of social capital put forth in Putnam (2000). Civic organizations such as clubs, churches, and fraternal groups are the essential elements used to evaluate the presence, absence, and amount of social connectedness in a community. I compiled a list of local civic organizations for each county using phone book listings, the identification of local chapters for each organization listed in the appendix of Putnam (2000), and organizations covered in the local daily newspapers during the siting process. I divided the population of the county by the total number of civic organizations.

[4] An injustice frame offers the best demonstration of the framing (Gamson, 1992; McAdam, 1999). When activists frame a contentious issue as an injustice, this creates a powerful political perception among individuals that helps mobilize support for the movement. Many environmental justice scholars use the injustice frame to explain mobilization against community environmental harms (Aronson, 1997; McGurty, 1995; Novotny, 1995). To assess the pervasiveness of an injustice frame, I identified letters to the editor in the local papers expressing a message either that the state government or siting agencies were motivated by political rather than technical considerations or that the state government was not representing local interests.

Variable	Estimates (Standard error)	P > /z/
Constant	2.634484	0.000
	(0.1155709)	
LCV score	−0.448281	0.013
	(0.1810829)	
Population/civic organizations	0.0003505	0.171
	(0.0002561)	
Injustice frame	4.861064	0.000
	(0.4825616)	
$N = 21$		
Chi square	97.98	0.0000
Pseudo $R^2 = 0.3142$		

Table A-2. Poisson Regression on the Number of Collective Acts of Opposition to the LLRW Facility in the 100 Days Following the First Opposition Event in Each Candidate County (Model 1: Classic Social Movement Agenda)

hypothesized that the higher the LCV score (that is, the more environmentally supportive the U.S. representative), the more actively opposed the county. But these results show just the opposite. The meaning of this finding is unclear. Perhaps communities are more likely to actively express their opposition to an LLRW site when they perceive a lack of environmental leadership from their representatives. These findings may also indicate that the LCV score is an invalid measure of preexisting political opportunities for LLRW activists. The prevalence of an injustice frame is significant and positively related to active opposition, in accordance with the classic social movement hypothesis.

VARIATIONS IN DISRUPTIVE PROTEST AND GOVERNMENT ACTIVITY

Table A.3 displays the results of event count analysis using Poisson regression, with the frequency of government-initiated collective acts of public opposition inversely related to the frequency of disruptive protest events. The measurement of each variable stems from local newspaper analysis over the course of the siting process in each of the candidate counties. Disruptive protest events were identified as those in which more than two individuals expressed opposition to the LLRW facility by directly confronting authority figures or

Variable	Estimates (Standard error)	$P > /z/$
Constant	1.148713	0.000
	(0.2473734)	
Government-initiated opposition events	−0.1207565	0.003
	(0.0412336)	
$N = 21$		
Chi square	9.85	0.002
Pseudo $R^2 = 0.10$		

Table A-3. Poisson Regression on the Number of Disruptive Protests in the First 100 Days of the Siting Process Following the First Opposition Event

sites of authority with marches, pickets, civil disobedience, or other noninstitutional activities. Government-initiated opposition events include any collective act of public opposition to the LLRW proposal initiated by a local government official or body. Even though these results are statistically significant and the pseudo R^2 indicates that the model explains some amount of variation, a small N and lack of variation in the dependent variable caution against relying heavily on this quantitative approach.

DEGREES OF IMPLEMENTATION PROGRESS

Table A.4 displays four bivariate ordinary least squares (OLS) regression models exploring the relationship between progress along the DOE milestones set out for implementation of the LLRWPA for construction of new LLRW disposal sites and collective acts of opposition to implementation. Model 1 measures all types of collective acts of public opposition in the first 100 days after the first such act for each candidate county. Model 2 measures only those acts that were disruptive, identified as protest events in which more than two individuals expressed opposition to the LLRW facility by directly confronting authority figures or sites of authority with marches, pickets, civil disobedience, or other noninstitutional activities. Model 3 measures only government-initiated acts of public opposition. Model 4 measures the first day during the siting process that local government expressed opposition to an LLRW site. The small N of this analysis based on the 21 candidate counties for an LLRW

Variable	Coefficient (Standard error)	Constant (Standard error) P > /t/	P > /t/	Adj. R²	N
Model 1: Number of collective acts of public opposition	−0.0658897 (0.0213091)	4.902329 (0.5136748) 0.000	0.006	0.2997	21
Model 2: Disruptive collective acts of public opposition	−0.1481959 (0.1356886)	3.880155 (0.4321389) 0.000	0.288	0.0096	21
Model 3: Government-initiated acts of public opposition	−0.166629 (0.0735166)	4.698169 (0.5788706) 0.000	0.035	0.1714	21
Model 4: Day of local government opposition	0.002129 (0.0006453)	3.039059 (0.3441883) 0.000	0.004	0.3308	21

Table A-4. Bivariate OLS Regression Models on the DOE Milestone Met in the Implementation Process for LLRW Candidate Counties

site weakens any conclusions that can be drawn from the quantitative analysis alone. Nevertheless, the results do support more qualitative analysis of the types of acts of public opposition. Not surprisingly, the overall number of collective acts of opposition is significantly and inversely related to implementation progress. The more acts of public opposition, the less progress was made toward building an LLRW facility. However, when only disruptive acts are considered, there is no significant relationship. Yet both the frequency and timing of local government-initiated opposition (Models 3 and 4) are significantly related to a lack of implementation progress. According to this analysis, both early and frequent local government opposition thwarted implementation progress toward the construction of LLRW disposal sites.

REFERENCES

Aeppel, Timothy. 2009. Nuclear-Power Industry Enjoys Revival 30 Years after Accident. *Wall Street Journal,* March 29, A3.

Albrecht, Stan L. 1999. Nuclear Gridlock: The Nuclear-Waste-Disposal Industry Is Haunted by Failed Policy. *Forum for Applied Research and Public Policy* 14:96–102.

Allen, David W., James P. Lester, and Kelly M. Hill. 2001. *Environmental Justice in the United States: Myths and Realities.* Boulder, CO: Westview Press.

Andrian, Charles F., and David E. Apter. 1995. *Political Protest and Social Change.* New York: New York University Press.

AP (Associated Press). 1988. Some Counties Scrambling to Get off Nuke Dump List. *Charleston (IL) Times-Courier,* Jan. 7, 7.

———. 1990a. 1,200 Waste Protestors Participate in March. *Richmond County Daily Journal,* Jan. 23, 1 .

———. 1990b. Nelson Ready to Blow Whistle on Progress of Waste Dump. *Norfolk (NE) Daily News,* Nov. 9, 1.

———. 1990c. Nelson Seeks Orr Debate near Proposed Nuclear Waste Site. *Norfolk (NE) Daily News,* Aug. 6, 2.

———. 1990d. Report: Other Waste Site Candidates Dropped. *Adrian (MI) Daily Telegram,* Feb. 16, 1.

Aronson, Hal R. 1997. Constructing Racism into Resources: A Portrait and Analysis of the Environmental Justice Movement. Ph.D. diss., University of California–Santa Cruz.

Bachrach, P., and M. Baratz. 1962. The Two Faces of Power. *American Political Science Review* 56:947–952.

Baumgartner, Frank R., and Bryan D. Jones. 1993. *Agendas and Instability in American Politics.* Chicago: University of Chicago Press.

Beckhorn, Sue. 1995. Interview conducted by Tom Peterson. Alfred, NY: Alfred Collection Archives, Alfred University.

Bell, Charles R., Jay E. Boudreau, and John G. Davis. 2004. Is There a Solution to the Electrical Energy Challenges in the U.S.? A Vision. Los Alamos, NM: Los Alamos National Laboratory.

Bennington (VT) Banner. 2009. Texas Firm Wants Nuke Waste. April 18.

Bouser, Steve. 1989a. One of Four Radioactive Dumpsites. *Salisbury (NC) Post,* Nov. 8, 1.

———. 1989b. Trying Hard Not to Be Out-NIMBY'd. *Salisbury (NC) Post,* Nov. 11, 2.

Brigman, Sam. 1989. Boost the Raiders. *Richmond County Daily Journal,* Dec. 5, 5.

Bullard, Robert D. 1983. Solid Waste Sites and the Black Houston Community. *Sociological Inquiry* 53:273–288.

———. 1993. *Confronting Environmental Racism: Voices from the Grassroots.* Boston: South End Press.

———, ed. 1994. *Unequal Protection: Environmental Justice and Communities of Color.* San Francisco: Sierra Club Books.

Bundy, Roger M. 1989. Get the Facts on Radioactive Waste before You Protest. *St. Clair (MI) Times Herald,* Nov. 13, 4.

Burns, Michael E., and William H. Briner. 1988. Setting the Stage. In *Low-Level Radioactive Waste Regulation: Science, Politics and Fear,* edited by M. E. Burns. Chelsea, MI: Lewis Publishers, 1–62.

Business Week. 1979. A Dangerous Dearth of Radioactive Dump Sites. Nov. 19, 54.

Butler, Rodney. 1989. More Opposition. *Richmond County Daily Journal,* Nov. 28, 3.

Calvert Cliffs' Coordinating Committee, Inc. v. U.S. Atomic Energy Commission. 1971. C.A.D.C.

Campbell, Sally. 1995a. Interview conducted by Tom Peterson. Allegany, NY.

Campbell, Stuart. 1995b. Interview conducted by Tom Peterson. Allegany, NY.

Carter, Jimmy. 1978. State of the Union Address. *Congressional Quarterly Almanac* 13E.

———. 1980. Presidential Message and Fact Sheet of February 12, 1980. In *The Politics of Nuclear Waste,* edited by E. William Colglazier, Jr. New York: Pergamon Press.

Carter, Luther J. 1988. *Nuclear Imperatives and Public Trust: Dealing with Radioactive Waste.* Washington, DC: Resources for the Future.

Cerrell Associates. 1984. *Political Difficulties Facing Waste-to-Energy Conversion Plant Siting.* Los Angeles: California Waste Management Board.

Clarke, Jeanne Nienaber, and Andrea K. Gerlak. 1998. Environmental Racism in Southern Arizona? The Reality beneath the Rhetoric. In *Environmental Injustices, Political Struggles: Race, Class and the Environment,* edited by D. E. Camacho. Durham, NC: Duke University Press, 82–100.

Coates, Dennis, Victoria Heid, and Michael Munger. 1994. Not Equitable, Not Efficient: U.S. Policy on Low-Level Radioactive Waste Disposal. *Journal of Policy Analysis and Management* 13 (3): 526–538.

Coch, William. 1995. Interview conducted by Tom Peterson. Alfred, NY: Alfred Collection Archives, Alfred University.

Condon, L. David. 1990. The Never Ending Story: Low-Level Waste and Exclusionary Authority of Noncompacting States. *Natural Resources Journal* 30: 65–86.

Condon, Lee. 1991a. Jackson Begins Cross-State Trek. *Manchester (CT) Journal Inquirer,* Aug. 12, 4.

———. 1991b. Residents Organize Grass-Roots Opposition. *Manchester (CT) Journal Inquirer,* June 11, 4.

Congressional Quarterly Almanac. 1954. Expansion of Atomic Energy Program.

———. 1973. AEC Authorization, 594–595.

———. 1976. Energy, 92.

———. 1979. What Happened at Three Mile Island?, 694.

Congressional Record. 1980a. *1980 Nuclear Waste Policy Act.* Washington, DC: U.S. Government Printing Office.

———. 1980b. Title II-Low-Level Radioactive Waste. Washington, DC: U.S. Government Printing Office.

———. 1980c. To Establish Repositories for Transuranic Waste, High Level Radioactive Waste, and Spent Fuel. Washington, DC: U.S. Government Printing Office.

Conlon, Kevin. 1988. County Opposes Radioactive Waste Dump Here. *Cortland (NY) Standard,* Dec. 29, 3.

———. 1989a. Cincy Meets on Dump. *Cortland (NY) Standard,* Jan. 3, 3.

———. 1989b. N-Dump Adds Foes. *Cortland (NY) Standard,* Jan. 11, 3.

Cook, Ruby. 1990. Radioactive Handouts. *Charleston (IL) Times-Courier,* March 28, 4.

Cortland (NY) Standard. 1989. Very Misleading Term, Sept. 26, 6.

Crum, Ron, Jason Sheely, and Raquel Jumonville. 1999. *Insights: Environmental Justice—A New Era in Environmental Permitting!.* Baton Rouge, LA: Radion International.

Cuomo, Mario. 1990. Statement by Governor Mario M. Cuomo. Albany, NY: New York State Executive Chamber Press Office, April 6, 1990.

———. 1996. Interview conducted by Tom Peterson. Alfred, NY.

Dahl, Robert. 1969. The Concept of Power. In *Political Power: A Reader in Theory and Research*, edited by Roderick Bell and David M. Edwards, R. Harrison Wagner. New York: Free Press, 36–41.

———. 1989. *Democracy and Its Critics*. New Haven, CT: Yale University Press.

Dalton Russell J., and Paula Garb, Nicholas P. Lovrich, John C. Pierce, John M. Whiteley, eds. 1999. *Critical Masses: Citizens, Nuclear Weapons Production, and Environmental Destruction in the United States and Russia*. Cambridge, MA: MIT Press.

Darling, Frances. 1989. Stop Nuclear Power. *Cortland (NY) Standard*, April 25, 7.

Dawkins, Don. 1989. Dawkins against Dump. *Richmond County Daily Journal*, Dec. 21, 7.

Dear, Michael. 1992. Understanding and Overcoming the NIMBY Syndrome. *Journal of the American Planning Association* 5892 (3): 288–306.

Della Porta, Donatella, and Mario Diani. 1999. *Social Movements: An Introduction*. Oxford: Blackwell.

Della Porta, Donatella, and Herbert Reiter. 1998. Introduction: The Policing of Protest in Western Democracies. In *Policing Protest: The Control of Mass Demonstrations in Western Democracies*, edited by Donatella della Porta and Herbert Reiter. Minneapolis: University of Minnesota Press, 1–32.

Denton, Van. 1989. Environmental Activists Gaining Clout in State. *Raleigh (NC) News & Observer*, Nov. 17, 1.

Dickenson, Joan. 1988. Local Reaction Cautious but Negative. *Olean (NY) Times Herald*, Dec. 21, 6.

———. 1989a. Citizens Seek County Funding to Help Fight Nuclear Dump. *Olean (NY) Times Herald*, Feb. 3, 11.

———. 1989b. Legislators Oppose Nuclear Dump Site. *Olean (NY) Times Herald*, Jan. 24, 10.

———. 1989c. West Almond Plan to Fight Nuclear Dump "Not Possible." *Olean (NY) Times Herald*, Jan. 25, 3.

———. 1989d. Whole County Has to Band Together. *Olean (NY) Times Herald*, Jan. 6, 10.

———. 1990. Protest Leaders Treat People "like Mushrooms." *Olean (NY) Times Herald*, Jan. 19, 11.

DOE (U.S. Department of Energy). 1992. *1991 Annual Report on Low-Level Radioactive Waste Management Progress*. Washington, DC: DOE.

————. 1995. *1994 Annual Report on Low-Level Radioactive Waste Management Progress.* Washington, DC: DOE.

————. 1996. *1995 Annual Report on Low-Level Radioactive Waste Management Progress.* Washington, DC: DOE.

Duffy, Robert J. 1997. *Nuclear Politics in America.* Lawrence, KS: University Press of Kansas.

Dunn, Ruby. 1990. Letter to the editor. *Richmond County Daily Journal,* Feb. 6, 5.

Eckstein, Rick. 1997. *Nuclear Power and Social Power.* Philadelphia: Temple University Press.

EIA (Energy Information Administration). 2001. *Annual Energy Review 2001.* Washington, DC: EIA.

Eisenhower, Dwight D. 1953. Atoms for Peace. www.iaea.org/About/history_speech.html (accessed August 20, 2010).

Eisinger, Peter K. 1973. The Conditions of Protest Behavior in American Cities. *American Political Science Review* 67 (1): 11–28.

Elliott, Janet. 2008. Company Wins License to Bury Nuclear Waste: Business Owned by a Top Perry Donor Will Dispose of "By-products" at West Texas Site. *Houston Chronicle,* May 22, A2.

Elster, Jon. 1998. A Plea for Mechanisms. In *Social Mechanisms: An Analytical Approach to Social Theory,* edited by P. Hedstrom and R. Swedberg. New York: Cambridge University Press, 45–73.

English, Mary R. 1992. *Siting Low-Level Radioactive Waste Disposal Facilities.* New York: Quorum Books.

EPA (U.S. Environmental Protection Agency). 1979. *Siting of Hazardous Waste Facilities and Public Opposition.* Washington, DC: SW-809.

————. 1998. *Final Guidance for Incorporating Environmental Justice Concerns in EPA's NEPA Compliance Analyses.* Washington, DC: EPA.

Epley. 1989. *Public Relations Assessment.* Raleigh, NC: Epley Associates.

Fahys, Judy. 2006. Utah's N-Waste Giant May Go Hotter. *Salt Lake Tribune,* Feb. 8, A1.

————. 2008. Regulator Says Utah Can't Be Only Solution for Nuclear Waste. *Salt Lake Tribune,* Sept. 3, A1.

————. 2009. Utah, EnergySolutions Square Off in Court. *Salt Lake Tribune,* Feb. 26, A1.

Farrell, Brian. 1988. Institutional and Regulatory Issues: A Utility Perspective on BRC *Proceedings of the Tenth Annual DOE Low-Level Radioactive Waste Management Conference.* August 30–September 1, 1988, Denver.

Farren, David. 1992. *Report on the Site Selection Process for the North Carolina LLRW Facility.* Raleigh, NC: Chatham County Board of Commissioners.

Fischer, Frank. 2000. *Citizens, Experts and the Environment: The Politics of Local Knowledge.* Durham, NC: Duke University Press.

Folsom, Geoff. 2008. Waste Control Specialists: Radioactive Site Looks to Expand. *Seattle-Tacoma (WA) Tribune Business News,* May 5.

Ford, Daniel. 1982. *The Cult of the Atom.* New York: Simon and Schuster.

Foreman, Christopher H. 1998. *The Promise and Peril of Environmental Justice.* Washington, DC: Brookings Institute.

FORRCE (For Richmond County Environment). 1990. News from FORRCE. *Richmond County Daily Journal,* Feb. 27, 3.

Franklin, Walt. 1989–1990. *Personal Diary.* Alfred, NY: Alfred Collection Archives, Alfred University.

Frey, Bruno, and Felix Oberholzer-Gee. 1997. The Cost of Price Incentives: An Empirical Analysis of Motivation Crowding Out. *American Economic Review* 87 (4): 746–755.

Frownfelder, David. 1990. Riga Stands Alone ... Maybe. *Adrian (MI) Daily Telegram,* Feb. 17, 1.

Gamson, William A. 1975. *The Strategy of Social Protest.* Homewood, IL: Dorsey.

———. 1992. The Social Psychology of Collective Action. In *Frontiers in Social Movement Theory,* edited by A. D. Morris and C. M. Mueller. New Haven, CT: Yale University Press, 53–76.

GAO (U.S. General Accounting Office/Government Accountability Office). 1983. *Siting of Hazardous Waste Landfills and Their Correlation with Racial and Economic Status of Surrounding Communities.* Washington, DC: GAO.

———. 1992. *Nuclear Waste: Slow Progress Developing Low-Level Radioactive Waste Disposal Facilities.* Washington, DC: GAO.

———. 1999. *Low-Level Radioactive Wastes: States Are Not Developing Disposal Facilities.* Washington, DC: GAO.

———. 2004. *Low-Level Radioactive Waste: Disposal Availability Adequate in the Short Term, but Oversight Needed to Identify Any Future Shortfalls.* Washington, DC: GAO.

———. 2008. *Low-Level Radioactive Waste: Status of Disposal Availability in the United States and Other Countries.* Washington, DC: GAO.

Gardner, Mary. 1990. Siting Process Is Not Legitimate. *Olean (NY) Times Herald,* Jan. 27, 7.

Gaventa, John. 1980. *Power and Powerlessness: Quiescence and Rebellion in an Appalachian Valley.* Chicago: University of Illinois Press.

Gerrard, Michael B. 1994. *Whose Backyard, Whose Risk: Fear and Fairness in Toxic and Nuclear Waste Siting.* Cambridge, MA: MIT Press.

Gershey, Edward L., Robert C. Klein, Esmeralda Party, and Amy Wilkerson. 1990. *Low-Level Radioactive Waste: From Cradle to Grave*. New York: Van Nostrand Reinhold.

Gibson, Holland M. 1989. Opposed to the Dump. *Richmond County Daily Journal*, Dec. 12, 5.

Glaberson, William. 1988. Coping in the Age of "NIMBY." *New York Times*, June 19, 3:1.

Gofman, John W., and Arthur W. Tamplin. 1971. *Poisoned Power: The Case against Nuclear Power Plants*. Emmaus, PA: Rodale Press.

Goldman, Benjamin A., and Laura Fitton. 1994. *Toxic Waste and Race Revisited*. Washington, DC: Center for Policy Alternatives.

Greenhouse, Linda. 1992. Supreme Court Roundup; Justices Hear Attack on Waste Law. *New York Times*, March 31, A15.

Greenwood, Ted. 1982. Nuclear Waste Management in the United States. In *The Politics of Nuclear Waste*, edited by E. William Colglazier Jr. New York: Pergamon Press.

Griffen, Joy. 1989. And Still More. *Richmond County Daily Journal*, Nov. 28, 3.

Hanson, Russell L. 1998. The Interaction of State and Local Governments. In *Governing Partners: State-Local Relations in the United States*, edited by R. L. Hanson. Boulder, CO: Westview Press, 1–16.

Hasper, John. 1995. Interview conducted by Tom Peterson. Allegany, NY: August 30.

Hatch, Daniel. 1991. Three Sites Eyed for Nuke Waste Dump. *Manchester (CT) Journal Inquirer*, June 11, 3.

Hayden, F. Gregory, and Steven R. Bolduc. 1997. Political and Economic Analysis of Low-Level Radioactive Waste. *Journal of Economic Issues* 31 (June):605–613.

Health Physics Society. 2005. *Position Statement: Low-Level Radioactive Waste Management Needs a Complete and Coordinated Overhaul*. McLean, VA: Health Physics Society.

Hedström, Peter, and Richard Swedberg. 1998. Social Mechanisms: An Introductory Essay. In *Social Mechanisms: An Analytical Approach to Social Theory*, edited by Peter Hedstrom and Richard Swedberg. New York: Cambridge University Press, 1–31.

Hill, Jeffrey S., and Carol S. Weissert. 1995. Implementation and the Irony of Delegation: The Politics of Low-Level Radioactive Waste Disposal. *Journal of Politics* 57 (2): 344–369.

Hoffman, Ian. 1989a. Dump: Waste Fight Unifies Union. *Monroe (NC) Enquirer-Journal*, Nov. 15, 1.

———. 1989b. Protestors Applaud Commission's Plan of Attack. *Monroe (NC) Enquirer-Journal*, Nov. 21, 1.

Hogwood, Brian W., and Guy B. Peters. 1985. *The Pathology of Public Policy.* New York: Oxford University Press.

Holland, Jeff. 1989a. Ariail to Chair Group Opposed to Waste Site. *Richmond County Daily Journal,* Nov. 22, 1.

———. 1989b. Hamlet Council Says No to Low-Level Nuke Dump. *Richmond County Daily Journal,* Nov. 15, 1.

———. 1989c. Public Hearing Draws over 500 to Courthouse. *Richmond County Daily Journal,* Dec. 8, 1.

Hollingsworth, Jerry. 1990. Opposed to Dump. *Richmond County Daily Journal,* March 11, 7.

Hunold, Christian. 2002. Canada's Low-Level Radioactive Waste Disposal Problem: Voluntarism Reconsidered. *Environmental Politics* 11 (2): 49–72.

Hunter, Susan, and Kevin M. Leyden. 1995. Beyond NIMBY: Explaining Opposition to Hazardous Waste Facilities. *Policy Studies Journal* 23 (4): 601–618.

Illinois Special Counsel. 1990. *Report of Special Counsel to the Illinois Senate Executive Subcommittee on Siting a Low Level Radioactive Waste Facility.* Springfield, IL: Special Counsel to the Illinois Senate.

Inhaber, Herbert. 1998. *Slaying the NIMBY Dragon.* New Brunswick, NJ: Transaction.

Ivins, Molly. 1980. Report Urges Regional Plan on Radioactive Waste Sites. *New York Times,* Aug. 3, 1.

Jefferds, Margaret. 1990. Get Involved, Get Informed. *Olean (NY) Times Herald,* Jan. 26, 5.

Jones, Spike. 1995. Interview conducted by Tom Peterson. Alfred, NY: Alfred Collection Archives, Alfred University. September 17.

Kagan, Robert, A. 2001. *Adversarial Legalism: The American Way of Law.* Cambridge, MA: Harvard University Press.

Kearney, Richard C., and Ande A. Smith. 1994. The Low-Level Radioactive Waste Siting Process in Connecticut: Anatomy of a Failure. *Policy Studies Journal* 22 (4): 617–628.

Keillor, Garrison. 1998. *Wobegon Boy.* New York: Penguin.

Kelley, Richard. 1995. Interview conducted by Tom Peterson. Allegany, NY.

Kemeny Commission. 1979. *The Need for Change: The Legacy of TMI.* Report of the President's Commission on the Accident at Three Mile Island. Washington, DC: Government Printing Office.

Kemp, Ray. 1992. *The Politics of Radioactive Waste Disposal.* Manchester: University of Manchester Press.

King, Gary. 1988. Statistical Models for Political Science Event Counts: Bias in Conventional Procedures and Evidence for the Exponential Poisson Regression Model. *American Journal of Political Science* 32 (3): 838–863.

Kraft, Michael E., and Bruce B. Clary. 1993. Public Testimony in Nuclear Waste Repository Hearings: A Content Analysis. In *Public Reactions to Nuclear Waste: Citizens' Views of Repository Siting*, edited by R. E. Dunlap, M. E. Kraft, and E. A. Rosa. Durham, NC: Duke University Press, 89–114.

Krannich, Richard S., and Stan L. Albrecht. 1995. Opportunity/Threat Responses to Nuclear Waste Disposal Facilities. *Rural Sociology* 60:435–453.

Kunreuther, Howard, and Douglas Easterling. 1990. Are Risk-Benefit Tradeoffs Possible in Siting Hazardous Facilities? *American Economic Review* 80 (2): 252–256.

Lenton, Garry. 2008. Disposal of Nuclear Waste Nears Crisis Stage—Nation Running Out of Room for Material. *Harrisburg (PA) Patriot News*, June 9, A1.

Lesbirel, S. Hayden. 1998. *NIMBY Politics in Japan: Energy Politics and the Management of Environmental Conflict*. Ithaca, NY: Cornell University Press.

Lloyd, Gary. 1995. Interview conducted by Tom Peterson. Allegany, NY.

LLW (Low-Level Radioactive Waste) Forum. 2006. *LLW Forum Discussion of Issues Statement*. Adopted 9/22/05; amended 9/18/06. Washington, DC: LLW Forum.

Lober, Douglas J. 1995. Why Protest? Public Behavior and Attitudinal Responses to Siting a Waste Disposal Facility. *Policy Studies Journal* 23 (3): 499–513.

London, Terry. 1989. Panel Must Offer Research Case against Waste Site. *St. Clair (MI) Times Herald*, Nov. 8, 6A.

Lowi, Theodore, J. 1978. Europeanization of America? From United States to United State. In *Nationalizing Government: Public Policies in America*, edited by Theodore J. Lowi and Alan Stone. Beverly Hills, CA: Sage, 15–29.

———. 1985. The State in Politics: The Relation between Policy and Administration. In *Regulatory Policy and the Social Sciences*, edited by R. G. Noll. Berkeley: University of California Press, 61–104.

Lowi, Theodore J., Benjamin Ginsberg, Elliot J. Feldman, Gregory J. Nigosian, Jonathan Pool, Allan Rosenbaum, Carlyn Rottsolk, Margaret Stapleton, Judith Van Heric, Julia Vitullo-Martin, and Thomas Vittulo-Martin. 1976. *Poliscide*. New York: Macmillan Publishing Company.

Lowry, Robert C. 1998. All Hazardous Waste Politics Is Local: Grass-roots Advocacy and Public Participation in Siting and Cleanup Decisions. *Policy Studies Journal* 26 (4): 748–759.

Lucey, Jim. 1995. Interview conducted by Tom Peterson. Alfred, NY: Alfred Collection Archives, Alfred University.

MacCallum, Tom. 1990. Waste Issues Dominate Commission Meeting. *Richmond County Daily Journal*, Jan. 9, 1.

Mather, Tom, and Kim Brooks. 1989. Counties' Opposition to Waste Site Varies. *Raleigh (NC) News & Observer*, Nov. 9, 1.

Mayer, Kenneth R. 1995. Closing Military Bases (Finally): Solving Collective Dilemmas through Delegation. *Legislative Studies Quarterly* 20:393–413.

Mazuzan, George T., and J. Samuel Walker. 1984. *Controlling the Atom: The Beginnings of Nuclear Regulation, 1946–1962.* Los Angeles: University of California Press.

McAdam, Doug. 1996. Conceptual Origins, Current Problems, Future Directions. In *Comparative Perspectives on Social Movements: Political Opportunities, Mobilizing Structures, and Cultural Framing*, edited by D. McAdam, J. McCarthy, and M. Zald. New York: Cambridge University Press, 23–41.

———. 1999. *Political Process and the Development of Black Insurgency, 1930–1970.* 2nd ed. Chicago: University of Chicago Press.

McAdam, Doug, Sidney Tarrow, and Charles Tilly. 2001. *Dynamics of Contention.* New York: Cambridge University Press.

McAvoy, George. 1999. *Controlling Technocracy: Citizen Rationality and the NIMBY Syndrome.* Washington, DC: Georgetown University Press.

McCarthy, John D., and Mayer N. Zald. 1987. Resource Mobilization and Social Movements: A Partial Theory. In *Social Movements in an Organizational Society*, edited by J. D. McCarthy and M. N. Zald. New Brunswick: Transaction Books, 15–47.

McCutcheon, Chuck. 2002. *Nuclear Reactions: The Politics of Opening a Radioactive Waste Disposal Site.* Albuquerque: University of New Mexico Press.

McGurty, Eileen Maura. 1995. The Construction of Environmental Justice: Warren County, North Carolina. Ph.D. diss., University of Illinois–Urbana-Champaign.

McPhee, John. 1974. *The Curve of Binding Energy: A Journey into the Awesome and Alarming World of Theodore B. Taylor.* New York: Farrar, Straus and Giroux.

Mencken, Henry Louis. 1949. *A Mencken Chrestomathy.* New York: A. A. Knopf.

Meyer, David S., and Debra C. Minkoff. 1997. Operationalizing Political Opportunity. Paper presented at American Sociological Association. August 1997, Toronto, Canada.

Meyer, David S., and Sidney Tarrow. 1998. A Movement Society: Contentious Politics for a New Century. In *The Social Movement Society: Contentious Politics for a New Century*, edited by David S. Meyer and Sidney Tarrow. New York: Rowman and Littlefield, 1–28.

Miller, Joe H. 1989. Radioactive Waste Isn't Wanted Here. *Salisbury (NC) Post*, Dec. 10, 6.

Mohai, Paul, and Bunyon Bryant. 1992. Environmental Racism: Reviewing the Evidence. In *Race and the Incidence of Environmental Hazards: A Time for Discourse*, edited by B. Bryant and P. Mohai. Boulder, CO: Westview Press, 163–176.

Mohai, Paul, and Robin Saha. 2007. Racial Inequality in the Distribution of Hazardous Waste: A National-Level Reassessment. *Social Problems* 54:343–370.

Monroe (NC) Enquirer-Journal. 1989a. Fight Looms: Nuclear Wastes Dumped Here Not Economical. Nov. 10, 1.

————. 1989b. NIMBY's Cause May Be Our Own. Dec. 10, 4.

Morris, Aldon D. 1984. *The Origins of the Civil Rights Movement: Black Communities Organizing for Change.* New York: Free Press.

Morris, Marjorie. 1989. Waste Site News Draws Reaction from Residents. *Sanford (NC) Herald,* Nov. 10, 1.

Mostaghel, Deborah M. 1994. The Low-Level Radioactive Waste Policy Amendments Act: An Overview. *DePaul Law Review* 43:379–421.

Mt. Vernon (IL) Register-News. 1988. Landowners Oppose Nuclear Dump. Sept. 23, 4.

Munton, Don. 1996. Introduction: The NIMBY Phenomenon and Approaches to Facility Siting. In *Hazardous Waste Siting and Democratic Choice,* edited by Don Munton. Washington, DC: Georgetown University Press, 1–54.

Myers, Betsy. 1995a. Interview conducted by Tom Peterson. Allegany, NY.

Myers, Steve. 1995b. Interview conducted by Tom Peterson. Allegany, NY.

National Research Council. 1996. *Review of New York State Low-Level Radioactive Waste Siting Process.* Washington, DC: National Academies Press.

————. 2003. *Improving the Regulation and Management of Low-Activity Radioactive Waste. Interim Report on Current Regulations, Inventories, and Practices.* Washington, DC: National Academies Press.

NCSL (National Conference of State Legislatures). 1980. *Goals for State-Federal Action: Policy Resolutions of the National Conference of State Legislatures, 1980–1981.* Denver: NCSL.

Nettles, Vernon. 1990. People Not Powerless. *Richmond County Daily Journal,* Feb. 13, 5.

New York State. 1989–1990. *New York State Statistical Yearbook.* 16th ed. Albany, NY: Nelson A. Rockefeller Institute of Government.

New York v. United States of America. 1992. 505 U.S. 144.

NGA (National Governors' Association Energy and Natural Resources). 1980. *Low-Level Waste, a Program for Action: The Final Report of the National Governors' Association Task Force on Low-Level Radioactive Waste Disposal.* Washington, DC: NGA.

————. 2010. *Low-Level Radioactive Waste Disposal Policy Position. NR-19.* Washington, DC: NGA.

NIRS (Nuclear Information and Resource Service). 2009. United States Commercial "Low-Level" Radioactive Wastes Disposal Sites Fact Sheet. www.nirs.org/factsheets/fctsht.htm (accessed June 27, 2009).

Nogas, Connie. 1990. "Stinky" Tactics Continue. *Cortland (NY) Standard,* March 22, 3.

Nogas, Constance M. 1989. Dump Opponents Watch NY Reps. *Cortland (NY) Standard,* Sept. 18, 3.

Novotny, Patrick John. 1995. Framing and Political Movements: A Study of Four Cases from the Environmental Justice Movement. Ph.D. diss., University of Wisconsin–Madison.

NRC (Nuclear Regulatory Commission). 1997. *NRC News.* Washington, DC: NRC.

Nuclear News. 1991. Texas Judge Voids Choice of Fort Hancock Site. March, 78.

———. 1996. OU Study: Impacts of Barnwell Site Closure. April, 42.

NYSLLRWSC (New York State Low-Level Radioactive Waste Siting Commission). 1989. Careful Study to Identify Sites. *LLRW Frontline* (Spring): 1.

O'Gorman, Mark John. 1997. More than NIMBY: The Not-in-My-Backyard Syndrome and Community Responses to Controversial and Opposed Scientific and Technological (COST) Facility Siting Attempts in New York State. Ph.D. diss., Syracuse University, Syracuse, NY.

O'Hare, Michael, and Debra Sanderson. 1993. Facility Siting and Compensation: Lessons from the Massachusetts Experience. *Journal of Policy Analysis and Management* 12:364–376.

O'Toole, Thomas. 1979. National Labs Considered for Storage of A-Waste. *Washington Post,* Oct. 25, A17.

O'Toole, Thomas, and Bill Peterson. 1979. A-Wastes Rejected by South Carolina. *Washington Post,* April 12, A1.

Parker, Frank L. 1988. Low-Level Radioactive Waste Disposal. In *Low-Level Radioactive Waste Regulation: Science, Politics and Fear,* edited by M. E. Burns. Chelsea, MI: Lewis Publishers, 85–108.

Pelletier, Fleurette. 1995. Interview conducted by Tom Peterson. Alfred, NY: Alfred Collection Archives, Alfred University.

Pellow, David, and Robert Brulle. 2005. *Power, Justice and the Environment: A Critical Appraisal of the Environmental Justice Movement.* Cambridge: Massachusetts Institute of Technology Press.

Peterson, Bo. 2007. Senators Vow to Bar States from Nuke Site. *Charleston (SC) Post and Courier,* April 12, A1.

Peterson, Tom. 2002. *Linked Arms: A Rural Community Resists Nuclear Waste.* Albany, NY: SUNY Press.

Piven, Frances F., and Richard Cloward. 1979. *Poor People's Movements: Why They Succeed, How They Fail.* New York: Vintage Books.

Policelli, Eugene. 1991. Radioactive Waste Shouldn't Be Dumped in Anyone's Backyard. *Hartford (CT) Courant,* June 27.

Popper, Frank J. 1981. Siting LULUs. *Planning* (April): 12–15.

Portney, Kent E. 1991. *Siting Hazardous Waste Treatment Facilities: The NIMBY Syndrome*. New York: Auburn House.

Pulido, Laura. 1996. A Critical Review of the Methodology of Environmental Racism Research. *Antipode* 28 (2): 142–159.

Putnam, Robert. 2000. *Bowling Alone: The Collapse and Revival of American Community*. New York: Simon and Schuster.

Rabe, Barry G. 1994. *Beyond NIMBY: Hazardous Waste Siting in Canada and the United States*. Washington, DC: Brookings.

Rabe, Barry G., Jeremy Becker, and Ross Levine. 2000. Beyond Siting: Implementing Voluntary Hazardous Waste Siting Agreements in Canada. *American Review of Canadian Studies* 30 (4): 479–496.

Rabe, Barry G., and William C. Gunderson. 2008. Voluntarism and Its Limits: Canada's Search for Radioactive Waste-Siting Candidates. *Canadian Public Administration* 42 (2): 193–214.

Rabe, Barry G., William Gunderson, Hilary Frazer, and John M. Gillroy. 1994. NIMBY and Maybe: Conflict and Cooperation in the Siting of Low-Level Radioactive Waste Disposal Facilities in the United States and Canada. *Environmental Law* 24:67–122.

Rankin, William L., Stanley M. Nealey, and Barbara Desow Melber. 1984. Overview of National Attitudes toward Nuclear Energy: A Longitudinal Analysis. In *Public Reaction to Nuclear Power: Are There Critical Masses?* edited by William R. Freudenburg and Eugene A. Rosa. Boulder, CO: Westview Press, 41–68.

Rennie, Kevin, and Edward Graziani. 1991. Legislators Vow to Undo Decision on Radioactive Dump. *Manchester (CT) Journal Inquirer*, June 13, 49.

Resnikoff, Marvin. 1982. When Does Consultation Become Co-optation? When Does Information Become Propaganda? An Environmental Perspective. In *The Politics of Nuclear Waste*, edited by E. William Colglazier Jr. New York: Pergamon Press.

Richmond County Daily Journal. 1989a. Committee Meetings Open. Nov. 29, 1.

———. 1989b. Conder Vows to Fight Location of Dump Here. Nov. 9, 1.

———. 1989c. Health Dept. Says Waste Site Not Geologically Sound. Nov. 30, 1.

———. 1989d. Let's Protest with Dignity. Dec. 12, 4.

———. 1990. Opposes Sheriff. April 15, 7.

Riley, Richard W. 1979. Why South Carolina Said No. *Washington Post*, April 23, A23.

———. 1982. Foreword. In *The Politics of Nuclear Waste*, edited by E. William Colglazier Jr. New York: Pergamon Press.

Roberson, Lynn Earley. 1989a. Coalition to Oppose Waste Site. *Salisbury (NC) Post*, Nov. 14, 1.

———. 1989b. Cohen: "We're Fighting It with Every Drop of Blood." *Salisbury (NC) Post*, Nov. 9, 1.

————. 1989c. County to Play Hard Ball. *Salisbury (NC) Post*, Nov. 15, 1.

————. 1989d. Waste Site Tests to Begin. *Salisbury (NC) Post*, Nov. 9, 1.

Robertson, Bill. 1989. Against Waste Dump. *Richmond County Daily Journal*, Nov. 28, 3.

Roschke, Susan Holland. 1997. Cooperation for the Environment: A Comparative Perspective and Status Characteristic Theory Examination of the Environmental Justice Movement. Ph.D. diss., Cornell University.

Rosenberg, Gail. 1991. Chem-Nuke Says Coverage "Unfair." *Richmond County Daily Journal*, July 14, 5.

Scanlon, Michael. 1987. Town Still Wonders What Good There Could Be in Nuclear Waste. *El Paso Times*, Sept. 27, 1, 2.

Shankle, Larry. 1990. Letter to the editor. *Richmond County Daily Journal*, Feb. 16, 4.

Shevalier, Hazel. 1989. Give Back Our Freedom. *Cortland (NY) Standard*, Sept. 23, 6.

Shumway, J. Matthew, and Richard H. Jackson. 2008. Place Making, Hazardous Waste and the Development of Tooele County, Utah. *Geographical Review* 98 (4): 433–455.

Simmons, Gelther. 1989. Local Officials React with Anger. *Salisbury (NC) Post*, Nov. 9, 1.

Slovic, Paul. 1996. Perception of Risk from Radiation. *Radiation Protection Dosimetry* 68 (3/4): 165–180.

Smothers, Ronald. 1995. Waste Site Becomes a Toxic Battlefield. *New York Times*, Oct. 9, A10.

Snow, David, and Robert Benford. 1992. Master Frames and Cycles of Protest. In *Frontiers in Social Movement Theory*, edited by A. D. Morris and C. M. Mueller. New Haven, CT: Yale University Press, 133–155.

Snow, David, E. Burke Rochford, Steven Worden, and Robert Benford. 1986. Frame Alignment Processes, Micromobilization, and Movement Participation. *American Sociological Review* 51:464–481.

Snowden, Monica A. 1997. Low-Level Is Not Our Level: The Save Boyd County Association's Response to the Siting of a Low-Level Radioactive Waste Disposal Facility in Boyd County, Nebraska. Ph.D. diss., University of Nebraska.

Stare, Russell. 1990. Clark County Defeats Nuke Referendum. *Charleston (IL) Times-Courier* Nov. 7, 1.

Steinberg, Michael W. 1999. Making Sense of Environmental Justice. *Briefly: Perspectives on Legislation, Regulation and Litigation* 3 (6): 1–68.

Stephens, Mark. 1980. *Three Mile Island*. New York: Random House.

Stone, Deborah A. 1997. *Policy Paradox: The Art of Political Decision Making*. New York: W.W. Norton.

Summers, Craig, and Donald W. Hine. 1997. Nuclear Waste Goes on the Road: Risk Perceptions and Compensatory Tradeoffs in Single-Industry Communities. *Canadian Journal of Behavioural Science* 29 (3): 211–223.

Sumpter, Glen. 1989a. FORRCE Sets Date for Public Hearing on Waste Dump. *Richmond County Daily Journal,* Nov. 29, 1.

———. 1989b. Have a Happy Thanksgiving. *Richmond County Daily Journal,* Nov. 23, 14.

———. 1990. Commission Expected to Hire Lawyer to Fight Waste Dump. *Richmond County Daily Journal,* Feb. 22, 1.

Swicegood, Wayne. 1990. Need New Sheriff. *Richmond County Daily Journal,* Sept. 23, 7.

Szasz, Andrew. 1994. *Ecopopulism.* Minneapolis: University of Minnesota Press.

Szymanski, Gerry. 1989. Something Deadly. *Olean (NY) Times Herald,* Sept. 20, 7.

Tarrow, Sidney. 1994. *Power in Movement: Social Movements, Collective Action and Politics.* New York: Cambridge University Press.

———. 1998. *Power in Movement: Social Movements and Contentious Politics.* New York: Cambridge University Press.

Taylor, Ted. 1995. Interview conducted by Tom Peterson. Alfred, NY: Alfred Collection Archives, Alfred University.

Temples, James R. 1980. The Politics of Nuclear Power: A Sub-Government in Transition. *Political Science Quarterly* 95 (2): 239–260.

Thomas, Deborah Kern. 1989. Against the Dump. *Richmond County Daily Journal,* Dec. 26, 4.

Thomas, Larry Lee. 1993. Communication in the Public Sphere of a Community Conflict: The Case of Locating a Nuclear Waste Repository in Boyd County, Nebraska. Ph.D. diss., University of Nebraska, Lincoln.

Tilly, Charles. 1978. *From Mobilization to Revolution.* Reading, MA: Addison-Wesley.

———. 1995. Contentious Repertoires in Great Britain, 1758–1834. In *Repertoires and Cycles of Collective Action,* edited by M. Traugott. Durham, NC: Duke University Press, 15–42.

Timberlake, Bill. 1995. Interview conducted by Tom Peterson. Alfred, NY: Alfred Collection Archives, Alfred University.

Timney, Mary M. 1998. Environmental Injustices: Examples from Ohio. In *Environmental Injustices, Political Struggles: Race, Class and the Environment,* edited by D. E. Camacho. Durham, NC: Duke University Press, 179–193.

UCC (United Church of Christ, Commission on Racial Justice). 1987. *Toxic Wastes and Race in the United States.* New York: UCC.

U.S. Advisory Commission on Intergovernmental Relations. 1981. *In Brief: The Federal Role in the Federal System: The Dynamics of Growth*. Washington, DC: U.S. Government Printing Office.

U.S. Census Bureau. 2000. Selected Historical Decennial Census Population and Housing Counts. Census'90. www.census.gov/population/www/censusdata/hiscendata.html (accessed August 30, 2010).

U.S. Congress House Committee on Interior and Insular Affairs. 1980. *Atomic Energy Act Amendments of 1980.*

———. 1985. *Low-Level Radioactive Waste Policy Amendments Act of 1985.*

U.S. Congress House Subcommittee on Energy and the Environment, Committee on Interior and Insular Affairs. 1977. *Proposed Nuclear Waste Storage in Michigan.*

———. 1979a. *Nuclear Waste Management.*

———. 1980. *To Amend the Atomic Energy Act of 1954*. Washington, DC: U.S. Government Printing Office.

U.S. Congress House Subcommittee on Energy Research and Production, Committee on Science and Technology. 1979b. *Low-Level Nuclear Waste Burial Grounds.*

U.S. Congress Office of Technology Assessment. 1989. *Partnerships under Pressure: Managing Commercial Low-Level Radioactive Waste*. Washington, DC: Office of Technology Assessment.

U.S. Congress Senate Subcommittee on Energy Research and Development. 1985. *Low-Level Radioactive Waste Disposal.*

Vari, Anna, Patricia Reagan-Cirincione, and Jeryl L. Mumpower. 1994. *LLRW Disposal Facility Siting: Successes and Failures in Six Countries*. Boston: Kluwer Academic Publishers.

Vergakis, Brock. 2009. Utah Takes N-Waste from States with Dumps. *Salt Lake City Deseret News*, May 6, A1.

Visocki, Kathryn. 1988. *Incentives, Compensation and Other Magic Tricks: Will They Help in Establishing New Waste Disposal Sites?*. Raleigh, NC: Southeast Compact Commission for Low-Level Radioactive Waste Management.

Voorhis, Dan. 1989a. Dump Site Scares Lanes Creek Residents. *Monroe (NC) Enquirer-Journal*, Nov. 10, 1.

———. 1989b. Waste Dump: Lanes Creek Forms Group to Fight Site. *Monroe (NC) Enquirer-Journal*, Nov. 14, 1.

———. 1989c. Waste Site Plan Draws Fire. *Monroe (NC) Enquirer-Journal*, Nov. 9, 1.

Warren, Roland. 1996. Interview conducted by Tom Peterson. Alfred, NY: Alfred Collection Archives, Alfred University. May 31.

Weart, Spencer. 1988. *Nuclear Fear: A History of Images*. New York: Cambridge University Press.

Wees, Greg. 1989. Boyd County Waste Dump Voted Down. *Norfolk (NE) Daily News* Jan. 12, 1.

Weingart, John. 2007. *Waste Is a Terrible Thing to Mind: Risk, Radiation, and Distrust of Government.* New Brunswick, NJ: Rivergate Books.

Wilder, Forrest. 2008. Good to Glow. *Texas Observer* April 4.

———. 2009. Send Us Your Waste. *Texas Observer blog* www.texasobserver.org/article. php?aid=2979 (accessed June 27, 2009).

Williams, Bruce A., and Albert R. Matheny. 1995. *Democracy, Dialogue and Environmental Disputes: The Contested Languages of Social Regulation.* New Haven, CT: Yale University Press.

Wolsink, Maarten. 1994. Entanglement of Interests and Motives: Assumptions behind the NIMBY-Theory on Facility Siting. *Urban Studies* 31 (6): 851–867.

Zaccagni, Hope. 1995. Interview conducted by Tom Peterson. Alfred, NY: Alfred Collection Archives, Alfred University.

Zidko, Donna. 1989. Some Questions. *Norfolk (NE) Daily News* July 8, 4.

INDEX

Note: Page numbers in italics indicate figures and tables. Page numbers followed by an 'n' indicate notes.